ENGLISH HISTORY IN THE MAKING

VOLUME I

English History in the Making VOLUME I

READINGS FROM THE SOURCES, TO 1689

William L. Sachse UNIVERSITY OF WISCONSIN

XEROX

XEROX COLLEGE PUBLISHING

Lexington, Massachusetts · Toronto

Cover illustration and woodcuts by Robert and Marilyn Dustin.

COVER: Anonymous 17th century engraving of the coronation of Charles II. Courtesy of the Museum of Fine Arts, Boston, Harvey D. Parker Collection.

TITLE PAGE: Three lions designating the heraldry of all English kings.

PART I: Woodcut adapted from a 10th century calendar, showing the occupation of the particular month.

PART II: Adapted from the Bayeux Tapestry, depicting the death of Harold.

PART III: Woodcut of a knight, adapted from a detail of a 14th century stained glass window.

PART IV: Woodcut adapted from a portrait of Henry VIII.

PART V: Woodcut from a detail of an anonymous 17th century engraving, showing King Charles II.

ISB Number: 0-536-00495-1
Library of Congress Catalog Card Number: 67-10154
Printed in the United States of America.

PREFACE

The aim of this book is to provide college students with selections from the sources of English history from earliest times until the Glorious Revolution. A great variety of such materials exists, and much of it has been published. But often, interesting items must be searched out in such uncommon, unwieldly, and expensive works as to make them virtually inaccessible for classroom use. Various collections and samplings have been prepared over the years, but books of this sort tend to cover a narrow span of time or to deal with a special category of documents.

The source materials assembled here have been selected and arranged with the object of providing students with relevant and comprehensible illustrative material to supplement the treatment usually found in college textbooks. The period, covering seventeen centuries, has been chronologically divided into five parts. Within these major subdivisions, items are presented for the most part chronologically, but with such topical arrangement as to focus attention on major issues and developments – political, constitutional, economic, social, and intellectual. In recognition of the greater attention usually paid to the later centuries covered in these pages, about half the documents belong to the Tudor and Stuart eras.

In the selection of documents, I have, of course, found it necessary to include unmistakably basic items such as Magna Carta and the Bill of Rights, but I have endeavored to present readings from diverse sources – some of them relatively unfamiliar. Thus the collection includes statutes and state papers, administrative and parliamentary; manorial rolls and charters; chronicles and other historical pieces; law reports and decisions; memoirs and letters; tracts or larger works of a political, philosophical, religious, or literary character. It is my hope that the reader may come to appreciate how rich, varied, and interesting is this testimony of the past, and that he may explore it more fully elsewhere.

In the interest of readability, as well as space, I have in most instances presented only the more important and meaningful passages from documents, indicating all omissions in the customary fashion. In all cases, spelling and punctuation have been adjusted to modern usage. Unusual words or expressions are explained in footnotes. Each document is accompanied with information regarding its source. Where I have drawn on an existing translation or rendering of a difficult English text, I have indicated its location in an additional reference. My obligations in this regard are recorded in the table of acknowledgments.

William L. Sachse

ACKNOWLEDGMENTS

I am obliged to the following publishers for permission to publish extracts from the works cited. The location of these extracts in this book is indicated by the selection number following each citation.

George Allen and Unwin Ltd.

PEEL, ALBERT and LELAND H. CARLSON (eds.), *The Writings of Robert Harrison and Robert Browne* (1953) – No. 141.

G. Bell and Sons, Ltd.

BLAND, A. E., *The Normans in England, 1066–1154* (1914) – No. 29, 36, 43, 56.

BLAND, A. E., P. A. BROWN, and R. H. TAWNEY, *English Economic History: Select Documents* (1930) – No. 107, 112, 113.

ESDAILE, ARUNDELL, *The Age of Elizabeth, 1547–1603* (1912) – No. 140.

FORESTER, THOMAS, (trans.), *The Chronicle of Florence of Worcester* (1854) – No. 26.

HENDERSON, ERNEST F., *Select Historical Documents of the Middle Ages* (1896) – No. 60.

GILES, J. A. (ed.), *Roger of Wendover's Flowers of History* (1849) – No. 46, 49.

GILES, J. A. (ed.), *Six Old English Chronicles* (1872) – No. 3.

GILES, J. A. (ed.), *The Venerable Bede's Ecclesiastical History of England* (1878) – No. 5, 16, 17, 19.

RILEY, H. T. (ed.), *Ingulph's Chronicle of the Abbey of Croyland* (1854) – No. 105.

JONES, W. GARMAN, *York and Lancaster* (1914) – No. 101, 102.

LOCKE, A. A., *War and Misrule, 1307–99* (1913) – No. 84.

RILEY, H. T. (trans.), *Annals of Roger de Hoveden* (1853) – No. 58.

ROBIESON, W. D., *The Growth of Parliament and the War with Scotland, 1216–1307* (1914) – No. 68.

Basil Blackwell

HASSALL, W. O., *They Saw It Happen* (1957) – No. 98.

LASLETT, PETER (ed.), *Patriarcha and Other Political Works of Sir Robert Filmer* (1949) – No. 200.

University of California Press

McGarry, Daniel D. (ed.), *The Metalogicon of John of Salisbury* (1955) – No. 62.

Cambridge University Press

Colgrave, Bertram, *Life of Bishop Wilfrid by Eddius Stephanus* (1927) – No. 18.

Coulton, G. G., *Social Life in Britain from the Conquest to the Reformation* (1956) – No. 54, 55.

Robertson, A. J. (ed.), *Anglo-Saxon Charters* (1956) – No. 20.

Smith, Thomas, *De Republica Anglorum, a Discourse on the Commonwealth of England,* L. Alston (ed.), (1906) – No. 156, 159.

Tyndale, William, *The Obedience of a Christian Man* (Parker Society, 1848) – No. 119, 120.

The Clarendon Press

Armstrong, C. A. J. (ed.), *The Usurpation of Richard III* (1936) – No. 106.

Prothero, G. W. (ed.), *Select Statutes and Other Constitutional Documents Illustrative of the Reigns of Elizabeth and James I* (1906) – No. 159, 161, 162, 163.

Stubbs, William, *Select Charters and Other Illustrations of English Constitutional History* (1895) – No. 64.

Cornell University Press

Sabine, George W. (ed.), *The Works of Gerrard Winstanley.* Copyright 1941 by Cornell University. Used by permission of Cornell University Press – No. 186.

J. M. Dent and Sons, Ltd.

Sir John Froissart's Chronicles of England, France, Spain, etc. (Everyman's Library, 1911) – No. 70, 87.

Elyot, Thomas, *The Governour* (Everyman's Library, 1937) – No. 118.

Lloyd Thomas, J. M. (ed.), *The Autobiography of Richard Baxter* (n. d.) – No. 190.

E. P. Dutton and Co., Inc.

Works listed under J. M. Dent and Sons, Ltd.

Eyre and Spottiswoode (Publishers), Ltd.

Douglas, David C. (general ed.), *English Historical Documents.* Vol. I: Dorothy Whitelock (ed.), *English Historical Documents, c. 500–1042* (1955) – No. 8, 9, 13, 14, 23, 24. Vol. II: David C. Douglas and George

W. Greenaway (eds.), *English Historical Documents, 1042–1189* (1953) – No. 30, 31, 33, 37, 38, 42, 44, 57, 59.

Whitelock, Dorothy (ed.), *The Anglo-Saxon Chronicle* (1961) – No. 22, 34.

Ginn and Company

Cheyney, Edward P., *Readings in English History Drawn from the Original Sources* (1935) – No. 1, 2, 6, 10, 27, 28, 32, 35, 39, 40, 42, 45, 53, 69, 71, 72, 76, 88, 99, 100, 167.

Cook, Albert S. and Chauncey B. Tinker, *Select Translations from Old English Poetry* (1902) – No. 25.

Cook, Albert S. and Chauncey B. Tinker, *Select Translations from Old English Prose* (1908) – No. 21.

Hakluyt Society

Taylor, E. G. R. (ed.), *The Original Writings and Correspondence of the Two Richard Hakluyts* (1935) – No. 147.

Harper and Row, Publishers, Inc.

Stephenson, Carl and F. G. Marcham (eds. and trans.), *Sources of English Constitutional History.* Copyright 1937 by Harper and Brothers. Reprinted by permission of Harper and Row, Publishers – No. 7, 11, 12, 15, 41, 48, 50, 63, 64, 96, 97.

Harvard University Press

Abbott, Wilbur C., *The Writings and Speeches of Oliver Cromwell,* vol. III (1945) – No. 183.

Her Majesty's Stationery Office

By permission of the Controller.

Atkinson, Ernest G. (ed.), *Calendar of State Papers Relating to Ireland, 1599–1600* (1899) – No. 149.

Brewer, J. S. (ed.), *Letters and Papers Foreign and Domestic of the Reign of Henry VIII* (1872) – No. 122.

Firth, C. H. and R. S. Rait (eds.), *Acts and Ordinances of the Interregnum* (1911) – No. 181, 182.

Gayangos, de, Pascal (ed.), *Calendar of State Papers, Spanish* (1879) – No. 123.

Hume, Martin A. S. (ed.), *Calendar of Letters and State Papers, Spanish* (1892–1899) – No. 144.

Journals of the House of Commons

Journals of the House of Lords

Robertson, J. C. (ed.), *Materials for the History of Thomas Becket* (1875–1885) – No. 45.

Rotuli Parliamentorum

Statutes of the Realm

Hutchinson Educational Ltd.

MILLWARD, J. S., *Portraits and Documents: Sixteenth Century* (1961) – No. 135, 143.

Longmans, Green and Co. Limited

HENNINGS, MARGARET A., *England under Henry III* (1924) – No. 51, 52, 60, 61.

HUGHES, DOROTHY, *Illustrations of Chaucer's England* (1918) – No. 73, 77, 79, 80, 81, 83, 85, 89, 92.

JAMES, MARGARET and MAUREEN WEINSTOCK, *England during the Interregnum, 1642–1660* (1935) – No. 184.

STONE, THORA J., *England under the Restoration* (1923) – No. 193, 194.

WERNHAM, R. B. and J. C. WALKER, *England under Elizabeth, 1558–1603* (1932) – No. 145.

WILLIAMS, C. H., *England under the Early Tudors, 1485–1529* (1925) – No. 114.

The Macmillan Company

ADAMS, GEORGE B. and H. MORSE STEPHENS, *Select Documents of English Constitutional History* (copyright 1901) – No. 75, 91, 93.

GEE, HENRY and WILLIAM J. HARDY, *Documents Illustrative of English Church History* (copyright 1896) – No. 124.

Macmillan and Company, Ltd.

Works listed under The Macmillan Company.

Methuen and Co., Ltd.

DAVIES, R. TREVOR, *Documents Illustrating the History of Civilization in Medieval England, 1066–1500* (1926) – No. 47.

The Odyssey Press, Inc.

JONES, RICHARD FOSTER (ed.), *Francis Bacon: Essays, Advancement of Learning, New Atlantis, and Other Pieces.* Copyright 1937 by The Odyssey Press, Inc. Reprinted by permission of The Odyssey Press, Inc. – No. 211.

Oxford University Press, Inc.

DUNHAM, WILLIAM H. and STANLEY PARGELLIS, *Complaint and Reform in England, 1436–1714* (1938) – No. 104.

DOUGLAS, DAVID C. (general ed.), *English Historical Documents.* Vol. I: DOROTHY WHITELOCK (ed.), *English Historical Documents, c. 500–1042* (1955) – No. 8, 9, 13, 14, 23, 24. Vol. II: DAVID C. DOUGLAS and GEORGE

W. GREENAWAY (eds.), *English Historical Documents, 1042–1189* (1953) – No. 30, 31, 33, 37, 38, 42, 44, 57, 59.

University of Pennsylvania (Department of History)

HOWLAND, ARTHUR C. (ed.), *The Early Germans (Translations and Reprints from the Original Sources of European History,* Vol. VI, No. 3) – No. 4.

University of Pennsylvania Press

BURKE, ROBERT B. (trans.), *The Opus Majus of Roger Bacon* (1928) – No. 62.

A. and J. Picard & Cie.

COSNEAU, E., *Les Grands Traités de la Guerre de Cent Ans* (1889) – No. 72, 99.

G. P. Putnam's Sons

HUGHES, PAUL L. and ROBERT F. FRIES (eds.), *Crown and Parliament in Tudor-Stuart England* (1959) – No. 195.

Royal Historical Society

FIRTH, C. H. (ed.), *Selections from the Papers of William Clarke,* Vol. I (Camden Society, 1891) – No. 177.

NICHOLS, JOHN G. (ed.), *Chronicle of the Grey Friars of London* (Camden Society, 1852) – No. 130.

SCARGILL-BIRD, S. R. (ed.), *Custumals of Battle Abbey in the Reigns of Edward I and Edward II* (Camden Society, 1887) – No. 53.

WRIGHT, THOMAS (ed.), *Letters relating to the Suppression of the Monasteries* (Camden Society, 1843) – No. 128.

Rutgers University Press

WHITELOCK, DOROTHY (ed.), *The Anglo-Saxon Chronicle* (1961) – No. 22, 34.

Student Christian Movement Press Ltd.

Work listed under The Westminster Press.

The Westminster Press

FAIRWEATHER, EUGENE R. (ed.), *A Scholastic Miscellany: Anselm to Ockham,* Vol. X, *Library of Christian Classics.* Published 1956 by The Westminster Press. Used by permission. – No. 62.

CONTENTS

PART I ✤ FROM THE ROMAN TO THE NORMAN CONQUEST

PART III ✦ THE LATER MIDDLE AGES, 1272–1485

PART V ✤ The Stuart Era, 1603–1689

Part 1

FROM THE ROMAN TO THE NORMAN CONQUEST

SECTION A

Celtic and Roman Britain

***

1. Britain Described by Julius Caesar

Julius Caesar invaded Britain twice, in the years 55 and 54 B.C. His observations were set down in his "Commentaries" or memoirs. While probably not accurate in all respects, they give us invaluable information on the inhabitants of the southeastern part of the island as he saw them.

The inland portions of the island are inhabited by those who themselves say that according to tradition they are natives of the soil; the coast regions are peopled by those who crossed from Belgium for the purpose of making war. Almost all of these are called by the names of those states from which they are descended and from which they came hither. After they had waged war they remained there and began to cultivate the soil. The island has a large population, with many buildings constructed after the fashion of the Gauls, and abounds in flocks. For money they use either gold coins or bars of iron of a certain weight. Tin is found in the inland regions, iron on the seacoast; but the latter is not plentiful. They use imported bronze. All kinds of wood are found here, as in Gaul, except the beech and fir trees. They consider it contrary to divine law to eat the hare, the chicken, or the goose. They raise these, however, for their

---

SOURCE: *De Bello Gallico,* bk. v, chap. 12, 14; bk. iv, chap. 33, in Edward P. Cheyney, *Readings in English History Drawn from the Original Sources* (Boston: Ginn, 1935), 15–16.

own amusement and pleasure. The climate is more temperate than in Gaul, since there are fewer periods of cold. . . .

By far the most civilized are those who dwell in Kent. Their entire country borders on the sea, and they do not differ much from the Gauls in customs. Very many who dwell farther inland do not sow grain but live on milk and flesh, clothing themselves in skins. All the Britons paint themselves with woad,[1] which produces a dark blue color; and for this reason they are much more frightful in appearance in battle. They permit their hair to grow long, shaving all parts of the body except the head and the upper lip. Ten and twelve have wives common among them, especially brothers with brothers and parents with children; if any children are born they are considered as belonging to those men to whom the maiden was first married. . . .

This is their manner of fighting from chariots. At first the charioteers ride in all directions, usually throwing the ranks into confusion by the very terror caused by the horses, as well as by the noise of the wheels; then as soon as they have come between the squads of horsemen, they leap from the chariots and fight on foot. The drivers of the chariots then withdraw a little from the battle and place the chariots together, so that if the warriors are hard pressed by the number of the enemy, they have a safe retreat to their own. Their horsemen possess such activity and their foot soldiers such steadfastness in battle and they accomplish so much by daily training that on steep and even precipitous ground they are accustomed to check their excited horses, to control and turn them about quickly, to run out on the pole, to stand on the yoke, and then swiftly to return to the chariot.

2. The Romanization of Britain

The conquest of Britain by the Romans, effectively begun in A.D. 43, was in the course of around forty years extended as far as the lowlands of Scotland. In the meantime some progress was made in introducing the Roman way of life, as is seen in the following passage from the life of Agricola, Governor of Britain, 78–84, by his son-in-law, Tacitus, a Roman historian born about A.D. 55.

In order that these men [the natives] living far apart, unskilled, and eager for war might, by a taste of pleasure, become accustomed to peace

SOURCE: Tacitus, *Agricola,* chap. 21, in Edward P. Cheyney, *Readings in English History Drawn from the Original Sources* (Boston: Ginn, 1935), 27.

[1] A dye-producing plant, later superseded by indigo.

and quiet, he [Agricola] personally urged and publicly aided them to build temples, market places, and homes by assisting those who seemed so disposed, and by censuring the inactive; thus rivalry for honor took the place of compulsion. Further, he provided a liberal education for the sons of the chieftains and gave preference to the natural endowments of the Britons over the endeavors of the Gauls; so much so that those who recently were unfavorable to the Roman language were now eager for its literature. So our dress came to be held in honor, and the toga was often seen. Gradually they fell a prey to the allurements of vice, the porticoes, the baths, the dainties of the banquet; this in the judgment of the ignorant was called civilization, although it is really characteristic of slavery. . . .

3. The Decay of Roman Influence

The fifth and sixth centuries may be called the "dark ages" in British history, in the sense that there is almost no surviving documentary evidence to throw light on the period. Some idea of the disintegration of Britain as a Roman province, once it was subjected to the full impact of barbarian invasions, may be gained from the rather impressionistic account of Gildas (516?–570?), in his *De Excidio et Conquestu Britanniae* (Concerning the Destruction of Britain).

After this, Britain is left deprived of all her soldiery and armed bands, of her cruel governors, and of the flower of her youth, who went with Maximus,[2] but never again returned; and utterly ignorant as she was of the art of war, groaned in amazement for many years under the cruelty of two foreign nations — the Scots[3] from the northwest, and the Picts[4] from the north.

The Britons, impatient at the assaults of the Scots and Picts, their hostilities and dreadful oppressions, send ambassadors to Rome with letters entreating in piteous terms the assistance of an armed band to protect them, and offering loyal and ready submission to the authority of Rome, if they only would expel their invading foes. A legion is immediately sent . . . and provided sufficiently with arms. When they had crossed

SOURCE: Gildas, *De Excidio et Conquestu Britanniae,* sec. 14–20, in *Six Old English Chronicles,* J. A. Giles (ed.), (London: G. Bell and Sons, 1872), 305–08.

[2] Maximus headed an insurrection of the legions stationed in Britain, and became Emperor (383–388).
[3] The name given by the Romans to the inhabitants of Ireland. In the sixth century these people settled in northern Britain.
[4] A name applied by the Romans to inhabitants of Britain north of the Forth. It means "painted people."

over the sea and landed, they came at once to close conflict with their cruel enemies, and slew great numbers of them. All of them were driven beyond the borders, and the humiliated natives rescued from the bloody slavery which awaited them. . . .

The Roman legion had no sooner returned home in joy and triumph than their former foes, like hungry and ravening wolves, rushing with greedy jaws upon the fold which is left without a shepherd, and wafted both by the strength of oarsmen and the blowing wind, break through the boundaries and spread slaughter on every side, and like mowers cutting down the ripe corn, they cut up, tread underfoot and overrun the whole country.

And now again they send suppliant ambassadors, . . . imploring assistance from the Romans, . . . that their wretched country might not altogether be destroyed, and that the Roman name, which now was but an empty sound to fill the ear, might not become a reproach even to distant nations. Upon this the Romans, moved with compassion, as far as human nature can be, at the relation of such horrors, send forward, like eagles in their flight, their unexpected bands of cavalry by land and mariners by sea, and planting their terrible swords upon the shoulders of their enemies, they mow them down like leaves which fall at the destined period . . .

The Romans, therefore, left the country, giving notice that they could no longer be harassed by such laborious expeditions, nor suffer the Roman standards, with so large and brave an army, to be worn out by sea and land by fighting against these unwarlike, plundering vagabonds, but that the islanders, inuring themselves to warlike weapons, and bravely fighting, should valiantly protect their country, their property, wives and children, and, what is dearer than these, their liberty and lives . . .

No sooner were they gone than the Picts and Scots, like worms which in the heat of mid-day come forth from their holes, hastily land again from their canoes . . . Moreover, having heard of the departure of our friends, and their resolution never to return, they seized with greater boldness than before on all the country towards the extreme north . . .

Again, therefore, the wretched remnant, sending to Aetius,[5] a powerful Roman citizen, address him as follows: "To Aetius, now Consul for the third time: the groans of the Britons." And again . . . thus: "The barbarians drive us to the sea; the sea throws us back on the barbarians: thus two modes of death await us — we are either slain or drowned." The Romans, however, could not assist them, and in the meantime the discomfited people, wandering in the woods, began to feel the effects of a severe famine, which compelled many of them without delay to yield themselves up to their cruel persecutors, to obtain subsistence. Others of

[5] A noted Roman general (c. 396–454), who defeated Attila in 451.

them, however, lying hid in mountains, caves and woods, continually sallied out from thence to renew the war. . . .

SECTION B

The Germanic Invasions

4. The Germans Described by Tacitus

We may gain an approximate picture of the Anglo-Saxon invaders of Britain from the account of the Germans in Tacitus' *Germania.* Tacitus wrote three centuries before the great descent upon England, and knew little of the particular peoples that invaded Britain. However, it does not seem likely that the German way of life had changed very much over this period, or that the remote northern tribes were much different from other Germans.

I myself subscribe to the opinion of those who hold that the German tribes have never been contaminated by intermarriage with other nations, but have remained peculiar and unmixed and wholly unlike other peoples. Hence the bodily type is the same among them all, notwithstanding the extent of their population. They all have fierce blue eyes, reddish hair, and large bodies fit only for sudden exertion; they do not submit patiently to work and effort, and cannot endure thirst and heat at all, cold and hunger they are accustomed to because of their climate. . . .

They choose their kings on account of their ancestry, their generals for their valor. The kings do not have free and unlimited power, and the generals lead by example rather than command, winning great admiration if they are energetic and fight in plain sight in front of the line. But no one is allowed to put a culprit to death or to imprison him, or even to beat him with stripes except the priests, and then not by way of a punishment or at the command of the general, but as though ordered by the god who they believe aids them in their fighting. Certain figures and images taken from their sacred groves they carry into battle, but their greatest incitement to courage is that a division of horse or foot is

---

SOURCE: Tacitus, *Germania,* chap. 4–16, in *The Early Germans* (*Translations and Reprints from the Original Sources of European History*), Arthur C. Howland (ed.), (Philadelphia: University of Pennsylvania), *VI,* no. 3, 5–22.

not made up by chance or by accidental association, but is formed of families and clans; and their dear ones are close at hand, so that the wailings of the women and the crying of the children can be heard during the battle. These are for each warrior the most sacred witnesses of his bravery, these his dearest applauders. They carry their wounds to their mothers and their wives, nor do the latter fear to count their number and examine them while they bring them food and urge them to deeds of valor. . . .

Concerning minor matters the chiefs deliberate, but in important affairs all the people are consulted, although the subjects referred to the common people for judgment are discussed beforehand by the chiefs. Unless some sudden and unexpected event calls them together, they assemble on fixed days either at the new moon or the full moon, for they think these the most auspicious times to begin their undertakings. They do not reckon time by the number of days, as we do, but by the number of nights. So run their appointments, their contracts; the night introduces the day, so to speak. A disadvantage arises from their regard for liberty in that they do not come together at once as if commanded to attend, but two or three days are wasted by their delay in assembling. When the crowd is sufficient they take their places fully armed. Silence is proclaimed by the priests, who have on these occasions the right to keep order. Then the king or a chief addresses them, each being heard according to his age, noble blood, reputation in warfare and eloquence, though more because he has the power to persuade than the right to command. If an opinion is displeasing they reject it by shouting; if they agree to it they clash with their spears. The most complimentary form of assent is that which is expressed by means of their weapons. . . .

When they go into battle it is a disgrace for the chief to be outdone in deeds of valor and for the following not to match the courage of the chief; furthermore for any one of the followers to have survived his chief and come unharmed out of a battle is lifelong infamy and reproach. It is in accordance with their most sacred oath of allegiance to defend and protect him and to ascribe their bravest deeds to his renown. The chief fights for victory; the men of his following, for their chief. If the tribe to which they belong sinks into the lethargy of long peace and quiet, many of the noble youths voluntarily seek other tribes that are still carrying on war, because a quiet life is irksome to the Germans and they gain renown more readily in the midst of perils, while a large following is not to be provided for except by violence and war. For they look to the liberality of their chief for their war horse and their deadly and victorious spear; the feasts and entertainments, however, furnished them on a homely but liberal scale, fall to their lot as mere pay. The means for this bounty are acquired through war and plunder. Nor could you persuade them to till the soil and await the yearly produce so easily as you could induce them to stir up an enemy and earn glorious wounds. Nay, they even think it tame and stupid to acquire by their sweat what they can purchase by their blood.

In the intervals of peace they spend little time in hunting but much in idleness, given over to sleep and eating; all the bravest and most warlike doing nothing, while the hearth and home and the care of the fields is given over to the women, the old men, and the various infirm members of the family. The masters lie buried in sloth by that strange contradiction of nature that causes the same men to love indolence and hate peace. It is customary for the several tribesmen to present voluntary offerings of cattle and grain to the chiefs, which, though accepted as gifts of honor, also supply their wants. They are particularly delighted in the gifts of neighboring tribes, not only those sent by individuals, but those presented by states as such – choice horses, massive arms, embossed plates and armlets. We have now taught them to accept money also.

It is well known that none of the German tribes live in cities, nor even permit their dwellings to be closely joined to each other. They live separated and in various places, as a spring or a meadow or a grove strikes their fancy. They lay out their villages not as with us in connected or closely joined houses, but each one surrounds his dwelling with an open space, either as a protection against conflagration or because of their ignorance of the art of building. They do not even make use of rough stones or tiles. They use for all purposes undressed timber, giving no beauty or comfort. Some parts they plaster carefully with earth of such purity and brilliancy as to form a substitute for painting and designs in color. . . .

5. The Germanic Invasions

We have no contemporary description of the invasions of the Angles, Saxons and Jutes. Bede, Abbot of Jarrow, describes them in some detail in his *Historia Ecclesiastica* (Ecclesiastical History). Though he lived more than two centuries later, he demonstrates considerable ability in "co-ordinating the fragments of information which came to him through tradition, the relation of friends, or documentary evidence."[6]

SOURCE: J. A. Giles (ed.), *The Venerable Bede's Ecclesiastical History of England* (London: G. Bell and Sons, 1878), 23–25.

---

[6] Frank Stenton, *Anglo-Saxon England* (Oxford, The Clarendon Press, 1943), page 187.

In the year of our Lord 449, Martian being made Emperor with Valentinian,[7] and the forty-sixth from Augustus, ruled the Empire seven years. Then the nation of the Angles, or Saxons, being invited by the aforesaid king,[8] arrived in Britain with three long ships, and had a place assigned them to reside in by the same king, in the eastern part of the island, that they might thus appear to be fighting for their country, whilst their real intentions were to enslave it. Accordingly they engaged with the enemy, who were come from the north to give battle, and obtained the victory; which being known at home in their own country, as also the fertility of the country and the cowardice of the Britons, a more considerable fleet was quickly sent over, bringing a still greater number of men, which, being added to the former, made up an invincible army. The newcomers received of the Britons a place to inhabit, upon condition that they should wage war against their enemies for the peace and security of the country, whilst the Britons agreed to furnish them with pay.

Those who came over were of the three most powerful nations of Germany — Saxons, Angles, and Jutes. From the Jutes are descended the people of Kent, and of the Isle of Wight, and those also in the province of the West Saxons who are to this day called Jutes, seated opposite to the Isle of Wight. From the Saxons, that is, the country which is now called Old Saxony, came the East Saxons, the South Saxons, and the West Saxons. From the Angles, that is, the country which is called Anglia, and which is said, from that time, to remain desert to this day, between the provinces of the Jutes and the Saxons, are descended the East Angles, the Midland Angles, Mercians, all the race of the Northumbrians, that is, of those nations that dwell on the north side of the river Humber, and the other nations of the English. The two first commanders are said to have been Hengist and Horsa. . . .

In a short time swarms of the aforesaid nations came over into the island, and they began to increase so much that they became terrible to the natives themselves who had invited them. Then, having on a sudden entered into league with the Picts, whom they had by this time repelled by the force of their arms, they began to turn their weapons against their confederates. At first they obliged them to furnish a greater quantity of provisions; and, seeking an occasion to quarrel, protested that, unless more plentiful supplies were brought them, they would break the confederacy and ravage all the island; nor were they backward in putting their threats in execution. In short, the fire kindled by the hands of these pagans proved God's just revenge for the crimes of the people. . . .

For the barbarous conquerors . . . plundered all the neighboring cities and country, spread the conflagration from the eastern to the western sea,

[7] These two ruled jointly, 450–455.

[8] Vortigern, a British king; he invited them to repel the northern tribes.

without any opposition, and covered almost every part of the devoted island. Public as well as private structures were overturned; the priests were everywhere slain before the altars; the prelates and the people, without any respect of persons, were destroyed with fire and sword; nor were there any to bury those who had been thus cruelly slaughtered. Some of the miserable remainder, being captured in the mountains, were butchered in heaps. Others, spent with hunger, came forth and submitted themselves to the enemy for food, being destined to undergo perpetual servitude, if they were not killed even upon the spot. Some, with sorrowful hearts, fled beyond the seas. Others, continuing in their own country, led a miserable life among the woods, rocks, and mountains, with scarcely enough food to support life, and expecting every moment to be their last.

SECTION C

Anglo-Saxon England: the Ranks of Society

6. Rights and Duties of the Peasant

For many hundreds of years, the peasant tillers of the soil were the most numerous element in English society. In Anglo-Saxon times he was called the *gebur;* later he was known as the *villein.* From a document dated around the year 1000, usually called *Rectitudines Singularum Personarum* (Rights of Individual Persons), we have the following statement of his duties and rights.

The peasant's services are various: in some places burdensome, in others light or moderate. On some land he must work at week-work two days every week, at such work as he is required through the year . . . If he do carrying he is not required to work while his horse is out. He shall pay on Michaelmas day[9] 10d rent and on Martinmas day[10] 23 measures of barley and 2 hens; at Easter a young sheep or 2d, and he shall watch

---

SOURCE: Benjamin Thorpe, *Ancient Laws and Institutes* (London: Eyre and Spottiswoode, 1840), *I,* 434, in Edward P. Cheyney, *Readings in English History Drawn from the Original Sources* (Boston: Ginn, 1935), 73.

---

[9] September 29.
[10] November 11.

from Martinmas to Easter at his lord's fold as often as it is his turn. And from the time that they first plow, to Martinmas, he shall each week plow one acre and himself prepare the seed in his lord's barn. Also three acres extra work and two of grass. If he needs more grass, then he plows for it, as he is allowed. For his plowing rent he plows three acres and sows it from his own barn. And he pays his hearth-penny.[11] Two by two feed one hound, and each peasant gives six loaves to the swineherd when he drives his herd to pasture. On that land where this custom holds it pertains to the peasant that he shall have given to him for his outfit two oxen and one cow and six sheep and seven acres sown on his piece of land. After that year he must perform all services which pertain to him. And he must have given to him tools for his work, and utensils for his house. Then when he dies his lord takes back all.

The land law holds on some lands, but here and there, as I have said, it is heavier or lighter, for all land services are not alike. On some land the peasant must pay honey rent, on some meat rent, on some ale rent.[12] Let him who holds the shire take care that he knows what the old land-right is and what are the customs of the people.

7. The Oath of a Man to His Lord

The freeman class declined in the Anglo-Saxon centuries, as increased emphasis was placed on lordship. Apart from relationships growing out of economic dependency, we note governmental encouragement of the concept of lordship, as a means of preserving the peace and ordering society. By the tenth century it was decreed that every man should have a lord, and the following oath of a man to his lord had come into being:

By the Lord before whom this holy thing is holy, I will to N. be faithful and true, loving all that he loves and shunning all that he shuns, according to the law of God and the custom of the world; and never by will or by force, in word or in deed, will I do anything that is hateful to him; on condition that he will hold me as I deserve and will furnish all that was agreed between us when I bowed myself before him and submitted to his will.

SOURCE: F. Liebermann, *Die Gesetze der Angelsachsen* (Halle: Max Niemeyer, 1903–1916), *I*, 396, in Carl Stephenson and Frederick G. Marcham, *Sources of English Constitutional History* (New York: Harper and Row, 1937), 25.

---

11 A payment to the church by each free household on Holy Thursday.
12 I.e., payment in kind.

8. Social Gradations

Anglo-Saxon society tended to become increasingly stratified. In addition to a nobility of blood, there developed a class which gained privileges through service and wealth, known as the thegns. One indication of rank was the size of the wergild (man-money), the fine paid in cases of homicide to the relatives of the slain person. The following pronouncements have been attributed to Archbishop Wulfstan of York, and date from the early eleventh century.

Concerning Wergilds and Dignities

1. Once it used to be that people and rights went by dignities, and councillors of the people were then entitled to honor, each according to his rank, whether noble or ceorl,[13] retainer or lord.

2. And if a ceorl prospered, that he possessed fully five hides[14] of land of his own, a bell and a castle-gate, a seat and special office in the king's hall, then was he henceforth entitled to the rights of a thegn.

3. And the thegn who prospered, that he served the king and rode in his household band[15] on his missions, if he himself had a thegn who served him, possessing five hides on which he discharged the king's dues,[16] and who attended his lord in the king's hall, and had thrice gone on his errand to the king – then he was afterwards allowed to represent his lord with his preliminary oath, and legally obtain his [right to pursue a] charge, wherever he needed.

4. And he who had no such distinguished representative, swore in person to obtain his rights, or lost his case.

5. And if a thegn prospered, that he became an earl, then was he afterwards entitled to an earl's rights.

6. And if a trader prospered, that he crossed thrice the open sea at his own expense, he was then afterwards entitled to the rights of a thegn.

7. And if there were a scholar who prospered with his learning so that he took orders and served Christ, he should afterwards be entitled to so much more honor and protection as belonged by rights to that order, if he kept himself [chaste] as he should. . . .

SOURCE: F. Liebermann, *Die Gesetze der Angelsachsen* (Halle: Max Niemeyer, 1903–1916), *I*, 456, in Dorothy Whitelock (ed.), *English Historical Documents, c. 500–1042*, David C. Douglas (gen. ed.), (London: Eyre and Spottiswoode, 1955; New York: Oxford University Press), 432.

---

[13] The ordinary freeman.

[14] A measure varying, in accordance with local custom, from 30 to 120 acres.

[15] The king's bodyguard.

[16] I.e., military service as well as other public charges.

9. The Emancipation of Slaves

Slavery was not uncommon in the Anglo-Saxon era. But the number of slaves declined during the later years of the period, largely owing to the influence of the Christian Church, which encouraged emancipation. The following document gives evidence of the practice, as well as of the grim economic pressures which sometimes forced men to enslave themselves.

[Geatfleda] has given freedom for the love of God and for the need of her soul: namely Ecceard the smith and Aelfstan and his wife and all their offspring, born and unborn, and Arcil and Cole and Ecgferth [and] Ealdhun's daughter, and all those people whose heads she took for their food[17] in the evil days. Whosoever perverts this and robs her soul of this, may God Almighty rob him of this life and of the heavenly kingdom, and may he be accursed dead and alive ever into eternity. And also she has freed the man whom she begged from Cwaespatric, namely Aelfwold and Colbrand and Aelfsige and his son Gamel, Ethelred Tredewude and his stepson Uhtred, Aculf and Thurkil and Aelfsige. Whoever deprives them of this, may God Almighty and St. Cuthbert[18] be angry with them.

10. The Life of the People

Unfortunately, it is not easy to find contemporary documents describing in any very meaningful fashion the daily life of the common people of Anglo-Saxon England. Here and there glimpses may be caught, however, as in this translation of a dialogue written by Aelfric about A.D. 990, as an exercise by which students could be taught Latin by references to conditions familiar to them, set forth in both Latin and Anglo-Saxon.

PUPILS. We beg you, O master, to teach us to speak Latin correctly, for we are unlearned and speak ignorantly.

SOURCES: Benjamin Thorpe, *Diplomatarium Anglicum, Aevi Saxonici* (London: Macmillan, 1865), 621, in Dorothy Whitelock (ed.), *English Historical Documents, c. 500–1042,* David C. Douglas (gen. ed.), (London: Eyre and Spottiswoode, 1955; New York: Oxford University Press), 563–64.

Louis F. Klipstein (ed.), "Colloquium Aelfrici," *Analecta Anglo-Saxonica* (New York: G. P. Putnam's Sons, 1849), *I,* 195f., in Edward P. Cheyney, *Readings in English History Drawn from the Original Sources* (Boston: Ginn, 1935), 69–72.

---

17 Who became her slaves for sustenance.
18 A noted British monk; Bishop of Lindisfarne, 685–687.

TEACHER. What do you want to speak about?

PUPILS. What do we care what we speak about, so long as our speech is correct and useful, not foolish or improper?

TEACHER. Do you wish to be flogged as you learn?

PUPILS. We had rather be flogged for the sake of learning than remain ignorant, but we know you are kind and will not want to inflict blows on us unless you are compelled to by us.

TEACHER [to one of the pupils]. I ask thee what thou wilt say to me? . . . What kind of work hast thou?

PUPIL. I have taken vows as a monk and sing every day with the brethren seven times, and I am busy with reading and singing; but nevertheless I would like meanwhile to learn to speak in the Latin tongue.

TEACHER. What do these thy companions know?

PUPIL. Some are plowmen, some shepherds, some oxherds, some huntsmen, some fishermen, some hawkers, some peddlers, some shoemakers, some salt dealers, some bakers.

TEACHER. What dost thou say, plowman? . . . How dost thou carry on thy work?

PLOWMAN. O Master, I work too hard; I go out at early dawn driving the oxen to the field, and yoke them to the plow. There is no winter so bitter that I venture to remain at home, for fear of my lord; then when the oxen are yoked and the share and colter fastened to the plow, through the whole day must I plow a complete acre or more.

TEACHER. Hast thou a companion?

PLOWMAN. I have a certain boy who drives the oxen with a goad, and who is now hoarse from cold and shouting.

TEACHER. What else dost thou do in a day?

PLOWMAN. Certainly I have more to do. I have to fill the mangers of the oxen with hay, and water them, and carry their dung outside.

TEACHER. Oh! oh! That is much labor.

PLOWMAN. Yes, the labor is great because I am not free.

TEACHER. What dost thou say, shepherd? . . . Hast thou any labor?

SHEPHERD. Indeed, I have. In the early morning I drive my sheep to their pasture, and I stand over them in heat and cold with a dog, so that the wolves shall not devour them, and I bring them back to their fold, and I milk them twice a day, and I move their folds besides, and I make cheese and butter, and I am faithful to my lord.

TEACHER. Oxherd, what dost thou do?

OXHERD. O Master, I do much labor. When the plowman unyokes the oxen I lead them to the pasture, and the whole night I stand over them watching on account of thieves, and then at break of day I take them to the plowman well fed and watered.

TEACHER. Is this man one of your companions?

PUPIL. Yes, he is.

[Teacher and huntsman converse]:

TEACHER. Dost thou know anything? I know one craft. What is that? I am a huntsman. Whose? The king's. How dost thou carry on thy craft? I set my nets and spread them in a suitable place, and I set on my dogs so that they shall drive the wild animals and cause them to run unawares into the nets, and I strangle them in the nets. Canst thou not hunt without nets? Yes, I am able to hunt without nets. How? I chase wild animals with swift dogs. What wild animals dost thou take principally? I take deer and boars, and does and goats, and sometimes hares. Hast thou been hunting to-day? I have not because it is Sunday, but yesterday I was hunting. What didst thou capture? Two stags and one boar. How didst thou catch them? The stags I took in nets and the boar I stuck in the throat. How didst thou dare to stick the boar? The dogs drove him to me and I, standing in the way, suddenly stuck him. Thou wast certainly bold then. A huntsman must not be timid, for many kinds of wild beast live in the woods. What dost thou do with what thou dost capture? I give to the king whatsoever I capture because I am his huntsman. What does he give thee? He feeds and clothes me well, and sometimes he gives me a horse or a bracelet that I may carry on my work more cheerfully.

[Teacher and merchant converse]:

TEACHER. What dost thou say, merchant?

MERCHANT. I say that I am useful to the king, to ealdorman,[19] to rich men, and to the whole people. Why? I go aboard my vessel with my merchandise and over the seas and sell my things and buy precious wares which are not produced in this country, and I bring them hither to you with great danger over the seas; and sometimes I suffer shipwreck, scarcely escaping alive after the loss of all my possessions. What things dost thou bring to us? Purple dye and silk, precious gems and gold, different colored cloths and paints, wine and oil, ivory and brass, copper and tin, sulphur and glass, and things such as these. Art thou willing to sell thy things here just as thou hast bought them there? I am not willing. Of what advantage would my labor then be to me? I want to sell them here more dear than I bought them there, so that I can make some gain to support myself, my wife, and my son.

TEACHER. Thou, shoemaker, what dost thou do in thy work that is useful to us?

SHOEMAKER. My art is indeed very useful and indeed very necessary to you. How? I buy hides and skins and prepare them by my art and make of them leather goods of every kind: sandals and shoes, leggings and leather bags, bridles and collars, wallets and sacks; and no one of you is willing to pass the winter without my art.

[19] The principal royal official in the shire.

TEACHER. Salt dealer, of what use is thy trade to us?

SALT DEALER. My trade is of very great use to all of you. No one of you takes any enjoyment in his dinner or his supper unless my craft is friendly with him.

TEACHER. How is that?

SALT DEALER. What man enjoys pleasant foods without the savor of salt? Who fills his cellars or storehouse without my craft? Behold, all your butter and cheese would spoil unless I acted as your guardians over them. You would not even eat your vegetables without me.

The wood carver says: Which of you does not make use of my craft, since I construct houses and utensils and ships for all of you?

The smith replies: O wood carver, why dost thou talk that way when thou canst not bore a single hole without my craft?

The good adviser says: O comrades and good workmen, let us give up these contentions and let there be peace and concord between us, and let each be of service to the other in his trade, and let us all meet at the plowman's, where we shall have food for ourselves and fodder for our horses; and I give this advice to all workmen, that each one exercise his trade diligently, because he who deserts his trade will be deserted by his trade. Whether thou art a priest or a monk, or a layman or a soldier, busy thyself about it; be what thou art, because it is a great loss and shame to a man not to be willing to be that which he is and ought to be.

SECTION D

Anglo-Saxon England: Government and Law

11. Allegiance to the King

The concept of a binding allegiance on the part of all men, including those with no personal ties to the king, developed slowly in Anglo-Saxon England. Not until the time of Edmund (939–946) is this concept expressed in unmistakable fashion. It is to be found among the dooms [laws] of that monarch, in these words.

SOURCE: F. Liebermann, *Die Gesetze der Angelsachsen* (Halle: Max Niemeyer, 1903–1916), *I*, 190, in Carl Stephenson and Frederick G. Marcham, *Sources of English Constitutional History* (New York: Harper and Row, 1937), 17.

In the first place [he commands] that all, in the name of God before whom this holy thing is holy, shall swear fealty to King Edmund, as a man should be faithful to his lord, without dissension or betrayal, both in public and in secret, loving what he loves and shunning what he shuns; and from the day on which this oath is sworn that no one shall conceal [the breach of] this [obligation] on the part of a brother or a relative any more than on the part of a stranger.

12. The Coronation Oath of Edgar, 973

The "oath of allegiance" from subjects had its counterpart in the oath taken by the king, in which recognition was made that the king had obligations toward them. It is first encountered at the time of the coronation of Edgar (959–975).

This writing has been copied, letter by letter, from the writing which Archbishop Dunstan[20] gave our lord at Kingston on the day that he was consecrated as king, forbidding him to make any promise save this, which at the bishop's bidding he laid on Christ's altar:

In the name of the Holy Trinity I promise three things to the Christian people my subjects: first, that God's Church and all Christian people of my realm shall enjoy true peace; second, that I forbid to all ranks of men robbery and all wrongful deeds; third, that I urge and command justice and mercy in all judgments, so that the gracious and compassionate God who lives and reigns may grant us all his everlasting mercy.

13. The Hundred Ordinance

The English kingdom was divided into shires, and the shires into hundreds. Each of these divisions were important in the conduct of government, but the hundred was more closely related to the daily lives of the common people. In each hundred there was an official

---

SOURCES: F. Liebermann, *Die Gesetze der Angelsachsen* (Halle: Max Niemeyer, 1903–1916), *I*, 214, 216, in Carl Stephenson and Frederick G. Marcham, *Sources of English Constitutional History* (New York: Harper and Row, 1937), 18.

F. Liebermann, *Die Gesetze der Angelsachsen* (Halle: Max Niemeyer, 1903–1916), *I*, 192, 194, in Dorothy Whitelock (ed.), *English Historical Documents, c. 500–1042,* David C. Douglas (gen. ed.), (London: Eyre and Spottiswoode, 1955; New York: Oxford University Press), 393.

---

20 The great cleric who became St. Dunstan (924–988).

[reeve] representing the king, and a court [moot] for the transaction of judicial and other business. The following ordinance, or code of laws, relating to the hundred dates from the middle years of the tenth century. Present-day scholarship suggests that we "regard this ordinance as part of a movement to secure some uniformity of administrative arrangements when, after the re-conquest of the Danelaw, the Kings of Wessex were ruling all England."[21]

This is the ordinance on how the hundred shall be held.

1. First, that they are to assemble every four weeks and each man is to do justice to another.

2. If the need is urgent, one is to inform the man in charge of the hundred, and he then the men over the tithings;[22] and all are to go forth, where God may guide them, that they may reach [the thief]. Justice is to be done on the thief as Edmund decreed previously.

2.1. And the value of the stolen property is to be given to him who owns the cattle, and the rest is to be divided into two, half for the hundred and half for the lord – except the men; and the lord is to succeed to the men.

3. And the man who neglects this and opposes the decision of the hundred – and afterwards that charge is proved against him – is to pay 30 pence to the hundred, and on a second occasion 60 pence, half to the hundred, half to the lord.

3.1. If he does it a third time, he is to pay half a pound; at the fourth time he is to forfeit all that he owns and be an outlaw, unless the king allows him [to remain in] the land.

4. And we decreed concerning strange cattle, that no one was to keep any, unless he have the witness of the man in charge of the hundred or of the man over the tithing; and he [the witness] is to be very trustworthy.

4.1. And unless he has one of them, he is not to be allowed to vouch to warranty.[23]

5. Further, we decreed, if one hundred follows up a trail into another hundred, that is to be made known to the man in charge of that hundred, and he is then to go with them.

5.1. If he neglects it, he is to pay 30 shillings to the king.

6. If anyone evades the law and flees, he who supported him in that injury is to pay the simple compensation.[24]

21 *English Historical Documents, c. 500–1042,* ed. Dorothy Whitelock (New York, Oxford, 1955), 393.

22 Organized groups of men, commonly ten in number.

23 Be surety for, or answer for.

24 The meaning appears to be that he who by his support makes it possible for another to commit the offense is to pay the value of the stolen goods.

6.1. And if he is accused of abetting his escape, he is to clear himself according as it is established in the district.

7. In the hundred, as in any other court, it is our will that in every suit the common law be enjoined, and a day appointed when it shall be carried out.

7.1. And he who fails to appear on the appointed day – unless it is through his lord's summons – is to pay 30 shillings compensation, and on a fixed day perform what he should have done before. . . .

14. The Regulation of the Blood Feud

The development of court-administered justice was slow. The practice of private vengeance was still widely current in the tenth century, as is seen from this code of King Edmund (939–946), regulating the blood feud. The governing authorities sought to encourage the acceptance of compensation by the relatives of the injured party. In the case of homicide, this was the wergild, a payment which varied in accordance with the rank of the individual. In addition, the concept that certain individuals and places were under the special protection of the King contributed to the extension of his influence in maintaining law and order; increasingly any criminal act was regarded as a breach of the "King's peace."

Prologue. King Edmund informs all people, both high and low, who are in his dominion, that I have been inquiring with the advice of my councillors, both ecclesiastical and lay, first of all how I could most advance Christianity.

Prol. 1. First, then, it seemed to us all most necessary that we should keep most firmly our peace and concord among ourselves throughout my domain.

Prol. 2. The illegal and manifold conflicts which take place among us distress me and all of us greatly. We decreed then:

1. If henceforth anyone slay a man, he is himself to bear the feud, unless he can with the aid of his friends within twelve months pay compensation at the full wergild, whatever class he [the man slain] may belong to.

1.1. If, however, the kindred abandons him, and is not willing to pay compensation for him, it is then my will that all that kindred is to be

---

SOURCE: F. Liebermann, *Die Gesetze der Angelsachsen* (Halle: Max Niemeyer, 1903–1916), *I*, 186, in Dorothy Whitelock (ed.), *English Historical Documents, c. 500–1042,* David C. Douglas (gen. ed.), (London: Eyre and Spottiswoode, 1955; New York: Oxford University Press), 391–92.

exempt from the feud, except the actual slayer, if they give him neither food nor protection afterwards.

1.2. If, however, any one of his kinsmen harbors him afterwards, he is to be liable to forfeit all that he owns to the king, and to bear the feud as regards the kindred [of the man slain], because they previously abandoned him.

1.3. If, however, anyone of the other kindred takes vengeance on any man other than the actual slayer, he is to incur the hostility of the king and all his friends, and to forfeit all that he owns.

2. If anyone flees to a church or my residence, and he is attacked or molested there, those who do it are to be liable to the same penalty as stated above.

3. And I do not wish that any fine for fighting or compensation to a lord for his man shall be remitted.

4. Further, I make it known that I will allow no resort to my court before he [the slayer] has undergone ecclesiastical penance and paid compensation to the kindred, [or] undertaken to pay it, and submitted to every legal obligation, as the bishop, in whose diocese it is, instructs him.

5. Further, I thank God and all of you who have well supported me, for the immunity from thefts which we now have; I now trust to you, that you will support this measure so much the better as the need is greater for all of us that it shall be observed.

6. Further, we have declared concerning *mundbryce*[25] and *hamsocn*,[26] that anyone who commits it after this is to forfeit all that he owns, and it is to be for the king to decide whether he may preserve his life.

7. Leading men must settle feuds: First, according to the common law the slayer must give a pledge to his advocate, and the advocate to the kinsmen, that the slayer is willing to pay compensation to the kindred.

7.1. Then afterwards it is fitting that a pledge be given to the slayer's advocate, that the slayer may approach under safe-conduct and himself pledge to pay the wergild.

7.2. When he has pledged this, he is to find surety for the wergild.

7.3. When that has been done, the king's *mund*[27] is to be established; 21 days from that day *healsfang*[28] is to be paid; 21 days from then the compensation to the lord for his man; 21 days from then the first instalment of the wergild.

[25] Violation of one's right of protection over others.

[26] Attack on a household.

[27] Protection; a subsequent violent act will thus be an offense against the king.

[28] That part of the wergild going to the nearest kin.

15. The Regulation of Ordeals

In judicial trials, the procedures most commonly employed by the Anglo-Saxons were compurgation, or oath-helping, in which one of the parties, usually the defendant, produced men who would support his charge or denial; and the ordeal, which in various forms involved an appeal to the supernatural, and which was closely supervised by a priest in an atmosphere of religious ceremonial. The following is a tenth-century law governing ordeals:

And with regard to the ordeal, according to the commands of God and of the archbishops and of all the bishops, we order that, as soon as the fire has been brought to heat the [iron or water for the] ordeal, no one shall come into the church except the priest and the man to be tried. And [if the ordeal is by iron] nine feet, according to the feet of the man to be tried, shall be measured from the [starting] post to the [final] mark. If, on the other hand, it is to be [ordeal by] water, that shall be heated until it becomes boiling hot, whether the kettle is of iron or of brass, of lead or of clay. And if the process is "single," the hand shall be plunged in for the stone up to the wrist; if it is "threefold," up to the elbow. And when the [water for the] ordeal is ready, two men from each party shall go in, and they shall agree that it is as hot as we have ordered.

Then an equal number of men from both parties shall go in and stand along the church on each side of the ordeal, and all of them shall be fasting and shall have held themselves from their wives during the previous night. And the priest shall sprinkle them all with holy water and give them to taste of the holy water; and he shall give them the Book to kiss and [shall make over them] the sign of Christ's cross. And no one shall build up the fire any more after the consecration has been begun, but the iron shall lie on the embers until the last collect.[29] Then it shall be laid on the [starting] post. And nothing else shall be said inside the church except a prayer to God Almighty that He disclose the fullness of truth. And [after the man has undergone the ordeal] his hand shall be [bound up and] sealed; and after the third day it shall be inspected to see whether, within the dealed[30] wrapping, it is foul or clean. And if anyone breaks these provisions, the ordeal shall be [counted] a failure for him, and he shall pay a fine of 120s to the king.

SOURCE: F. Liebermann, *Die Gesetze der Angelsachsen* (Halle: Max Niemeyer, 1903–1916), *I*, 386–87, in Carl Stephenson and Frederick G. Marcham, *Sources of English Constitutional History* (New York: Harper and Row, 1937), 16.

---

[29] Prayer.
[30] Divided, parted.

SECTION E

The Progress of Christianity

16. The Conversion of Kent, 597

Although Christianity had been introduced to Britain under the Romans, the German invaders had never been Christianized. Thus Britain at the end of the sixth century was overwhelmingly pagan. Under Pope Gregory I, known as "the Great," Augustine, a Benedictine, was sent in 597 with a band of monks to work for conversion. Their efforts were directed toward the conversion of Kent; here King Ethelbert, who had married a Christian, gave them a friendly reception and was subsequently converted. This missionary enterprise is described by Bede in his *Ecclesiastical History:*

Augustine, thus strengthened by the confirmation of the blessed father Gregory, returned to the work of the word of God, with the servants of Christ, and arrived in Britain. The powerful Ethelbert was at that time king of Kent. He had extended his dominions as far as the great river Humber, by which the Southern Saxons are divided from the Northern. On the east of Kent is the large Isle of Thanet . . . In this island landed the servant of our Lord, Augustine, and his companions, being, as is reported, nearly forty men.

They had, by order of the blessed Pope Gregory, taken interpreters of the nation of the Franks, and sending to Ethelbert, signified that they were come from Rome, and brought a joyful message, which most undoubtedly assured to all that took advantage of it everlasting joys in heaven, and a kingdom that would never end, with the living and true God. The King having heard this, ordered them to stay in that island where they had landed, and that they should be furnished with all the necessaries, till he heard of the Christian religion, having a Christian wife, of the royal family of the Franks, called Bertha, whom he had received from her parents upon condition that she should be permitted to practice her religion with the bishop Luidhard, who was sent with her to preserve her faith.

Some days after, the King came into the island and, sitting in the open air, ordered Augustine and his companions to be brought into his presence.

SOURCE: J. A. Giles (ed.), *The Venerable Bede's Ecclesiastical History of England* (London: G. Bell and Sons, 1878), 36–39.

For he had taken precaution that they should not come to him in any house, lest, according to an ancient superstition, if they practiced any magical arts, they might impose upon him, and so get the better of him. But they came furnished with divine, not with magic virtue, bearing a silver cross for their banner, and the image of our Lord and Saviour painted on a board; and singing the litany, they offered up their prayers to the Lord for the eternal salvation both of themselves and of those to whom they were come. When he had sat down, pursuant to the King's commands, and preached to him and his attendants there present the word of life, the King answered thus: "Your words and promises are very fair, but as they are new to us, and of uncertain import, I cannot approve of them so far as to forsake that which I have so long followed with the whole English nation. But because you are come from far into my kingdom, and, as I conceive, are desirous to impart to us those things which you believe to be true and most beneficial, we will not molest you, but give you favorable entertainment, and take care to supply you with your necessary sustenance; nor do we forbid you to preach and gain as many as you can to your religion." Accordingly he permitted them to reside in the city of Canterbury, which was the metropolis of all his dominions, and, pursuant to his promise, besides allowing them sustenance, did not refuse them liberty to preach. It is reported that as they drew near to the city, after their manner, with the holy cross and the image of our sovereign Lord and King, Jesus Christ, they in concert sang this litany: "We beseech thee, O Lord, in all thy mercy, that thy anger and wrath be turned away from this city, and from thy holy house, because we have sinned. Hallelujah!". . .

As soon as they entered the dwelling-place assigned to them, they began to imitate the course of life practised in the primitive church . . . ; in short several believed and were baptized, admiring the simplicity of their innocent life, and the sweetness of their heavenly doctrine.

There was on the east side of the city a church dedicated to the honor of St. Martin, built whilst the Romans were still in the island, wherein the Queen, who, as has been said before, was a Christian, used to pray. In this they first began to meet, to sing, to pray, to say mass, to preach, and to baptize, till the King, being converted to the faith, allowed them to preach openly and build or repair churches in all places.

17. The Conversion of Northern England

The permanent introduction of Christianity into northern England was effected by missionaries from Scotland toward the middle of the seventh

---

SOURCE: J. A. Giles (ed.), *The Venerable Bede's Ecclesiastical History of England* (London: G. Bell and Sons, 1878), 111–12.

century. Chief among them was St. Aidan (d. 651), who became the first bishop of Lindisfarne, near the present Berwick-on-Tweed. This Celtic observance of Christianity differed from the Roman practice, in baptismal procedure, the method of the tonsure, and the dating of Easter, as is touched on in this account from Bede's Ecclesiastical History:

Oswald,[31] as soon as he ascended the throne, being desirous that all his nation should receive the Christian faith, whereof he had found happy experience in vanquishing the barbarians, sent to the elders of the Scots, among whom himself and his followers, when in banishment, had received the sacrament of baptism, desiring they would send him a bishop, by whose instruction and ministry the English nation which he governed might be taught the advantages and receive the sacraments of the Christian faith. Nor were they slow in granting his request, but sent him Bishop Aidan, a man of singular meekness, piety and moderation, zealous in the cause of God, though not altogether according to knowledge; for he was wont to keep Easter Sunday according to the custom of his country, which we have before so often mentioned, from the fourteenth to the twentieth moon; the northern province of the Scots and all the nations of the Picts celebrating Easter then after that manner. . . .

On the arrival of the bishop, the King appointed him his episcopal see in the isle of Lindisfarne, as he desired. . . . The King also, humbly and willingly in all cases giving ear to his admonitions, industriously applied himself to build and extend the church of Christ in his kingdom; wherein, when the bishop, who was not skilful in the English tongue, preached the gospel, it was most delightful to see the King himself interpreting the word of God to his commanders and ministers, for he had perfectly learned the language of the Scots during his long banishment.[32]. . .

18. The Council of Whitby, 664

Important in the development of a unified Christian Church in Anglo-Saxon England was the Council or Synod of Whitby, held in 664.

SOURCE: Bertram Colgrave, *Life of Bishop Wilfrid by Eddius Stephanus* (London: Cambridge University Press, 1927), 20–23.

---

[31] King of Northumbria, 635–642.
[32] To Iona, an island in the Inner Hebrides. Here St. Columba founded a monastery in 563, which became an important missionary center.

Here adherents of the Celtic Christianity, brought to Northumbria by Aidan, and of the Roman Christianity introduced by Augustine met to debate the merits of their respective teachings. A priest of the time, Eddius Stephanus, gives us an account of the Council in a Life of Bishop Wilfrid, written in the early eighth century. The decision to adopt Roman usages served to link England with the dominant Christian movement of western Europe, and facilitated the unification of the English Church.

On a certain occasion in the days of Colman,[33] bishop of York and metropolitan, while Oswy and Alhfrith his son were reigning,[34] the abbots and priests and men of all ranks in the orders of the Church gathered together in a monastery called Whitby, in the presence of the holy mother and post pious nun Hilda, as well as of the kings and two bishops, namely Colman and Agilberht, to consider the question of the proper date for the keeping of Easter – whether in accordance with the British and Scottish manner and that of the whole of the northern district, Easter should be kept on the Sunday between the fourteenth day of the moon and the twenty-second, or whether the plan of the apostolic see was better, namely to celebrate Easter Sunday between the fifteenth day of the moon and the twenty-first. The opportunity was granted first of all to Bishop Colman, as was proper, to state his case in the presence of all. He boldly spoke in reply as follows: "Our fathers and their predecessors, plainly inspired by the Holy Spirit as was Columba,[35] ordained the celebration of Easter on the fourteenth day of the moon, if it was a Sunday, following the example of the Apostle and Evangelist John 'who leaned on the breast of the Lord at supper' and was called the friend of the Lord. He celebrated Easter on the fourteenth day of the moon and we, like his disciples Polycarp and others, celebrate it on his authority; we dare not change it, for our fathers' sake, nor do we wish to do so. I have expressed the opinion of our party, do you state yours."

Agilberht the foreign bishop and Agatho his priest bade St. Wilfrid, priest and abbot, with his persuasive eloquence explain in his own tongue the system of the Roman Church and of the apostolic see. With his customary humility he answered in these words: "This question has already been admirably investigated by the three hundred and eighteen most holy and learned fathers gathered together in Nicaea, a city of Bithynia. They fixed amongst other decisions upon a lunar cycle which recurs every nineteen years. This cycle never shows that Easter is to be

[33] St. Colman (d. 676), of Irish origin. Actually he was Bishop of Lindisfarne.
[34] Oswy was King of Northumbria, 642–670; his son was subking of Deira, a division of Northumbria.
[35] St. Columba (521–597), who undertook the conversion of Scotland to Christianity.

kept on the fourteenth day of the moon. This is the fixed rule of the apostolic see and of almost the whole world, and our fathers, after many decrees had been made, uttered these words: 'he who condemns any one of these let him be accursed.' "

Then, after St. Wilfrid the priest had finished his speech, King Oswy smilingly asked them all: "Tell me which is greater in the kingdom of heaven, Columba or the Apostle Peter?" The whole synod answered with one voice and one consent: "The Lord settled this when he declared: 'Thou art Peter and upon this rock I will build my Church and the gates of hell shall not prevail against it. And I will give thee the keys of the kingdom of heaven; and whatsoever thou shalt bind on earth shall be bound in heaven; and whatsoever thou shalt loose on earth shall be loosed in heaven.' "[36]

The king wisely replied: "He is the porter and keeps the keys. With him I will have no differences nor will I agree with those who have such, nor in any single particular will I gainsay his decisions so long as I live."

So Bishop Colman was told what he must do, should he reject the tonsure and the Easter rule for fear of his fellow-countrymen, namely he must retire and leave his see to be taken by another and a better man. Thus indeed he did.

19. The Canons of Archbishop Theodore, 673

The unification and more efficient organization of the English church was significantly furthered by Theodore of Tarsus, appointed Archbishop of Canterbury in 668, and the first such prelate to whom the entire church submitted. Under him episcopal rights and duties were better defined, and the foundation was laid for a settled system of parishes, which increasingly served in place of the ministrations of itinerant missionaries attached to monasteries. In 673 a church council or synod was held at Hertford, at which the English church for the first time acted as one body. At this assembly Theodore recommended ten canons, or ecclesiastical rules, as of prime importance. Bede, in his *Ecclesiastical History*, describes them as follows:

CHAPTER I. That we all in common keep the holy day of Easter on the Sunday after the fourteenth moon of the first month.

SOURCE: J. A. Giles (ed.), *The Venerable Bede's Ecclesiastical History of England* (London: G. Bell and Sons, 1878), 182-83.

---

[36] Matthew 16: 18, 19.

II. That no bishop intrude into the diocese of another, but be satisfied with the government of the people committed to him.

III. That it shall not be lawful for any bishop to trouble monasteries dedicated to God, nor to take anything forcibly from them.

IV. That monks do not remove from one place to another, that is, from monastery to monastery, unless with the consent of their own abbot; but that they continue in the obedience which they promised at the time of their conversion.

V. That no clergyman, forsaking his own bishop, shall wander about, or be anywhere entertained without letters of recommendations from his own prelate. But if he shall be once received, and will not return when invited, both the receiver and the person received be under excommunication.

VI. That bishops and clergymen, when travelling, shall be content with the hospitality that is afforded them; and that it be not lawful for them to exercise any priestly function without leave of the bishop in whose diocese they are.

VII. That a synod be assembled twice a year; but in regard that several causes obstruct the same, it was approved by all that we should meet on the first of August once a year, at the place called Clofeshoch.[37]

VIII. That no bishop, through ambition, shall set himself before another; but that they shall all observe the time and order of their consecration.

IX. It was generally set forth, that more bishops should be made as the number of believers increased; but this matter for the present was passed over.

X. Of marriages: that nothing be allowed but lawful wedlock; that none commit incest; no man quit his true wife unless, as the Gospel teaches, on account of fornication. And if any man shall put away his own wife, lawfully joined to him in matrimony, that he take no other, if he wishes to be a good Christian, but continue as he is, or else be reconciled to his own wife.

These chapters being thus treated of and defined by all, to the end that for the future no scandal of contention might arise from any of us, or that things be falsely set forth, it was thought fit that every one of us should, by subscribing his hand, confirm all the particulars so laid down. Which definitive judgment of ours I dictated to be written by Titullus, our notary. Done in the month and indiction aforesaid. Whosoever, therefore, shall presume in any way to oppose or infringe this decision, confirmed by our consent and by the subscription of our hands, according to the decree of the canons, must take notice that he is excluded from all sacerdotal functions and from our society. May the divine grace preserve us in safety, living in the unity of his holy church.

[37] An unidentified place.

20. The Endowment of a Monastery, 1023

Monasteries played an important role in the christianizing of Britain. The founding and endowing of these institutions went on through most of the Anglo-Saxon period and continued after the Norman Conquest, reaching a peak in the twelfth century. The following charter illustrates the piety of King Canute (1017–1035) and the means by which he provided for the monks of Canterbury in 1023.

In the name of the most high God and of our Lord the Saviour Christ. Our holy and our righteous fathers with true insistence and with frequent admonition remind us that we should dread and love untiringly with inward joyfulness of heart and with zeal in good works the Almighty God whom we love and believe in. Since he shall requite all our works on the Day of Judgment, according to each man's deserts, let us strive with the utmost endeavor of our hearts to follow him. Although we are laden with the burden of this mortal life and defiled with the transitory possessions of this world, yet we may purchase the eternal reward of the heavenly life with these crumbling riches, and therefore I, Cnut, by the grace of God King of England and of all the adjacent islands, lay the royal crown from my head with my own hands upon the altar of Christ in Canterbury for the benefit of the said monastery, and I grant to the said monastery for the support of the monks the haven at Sandwich and all the landing places and the water dues from both sides of the river from Pepperness to *Maercesfleote,* whoever owns the land, in such wise that when it is high tide and a ship is afloat, the officers of Christchurch shall receive the dues from as far inland as can be reached by a small axe thrown from the ship. And no one shall ever in any kind of way have any control in the said haven except the monks of Christchurch, and theirs shall be the ship and the ferrying across the haven and the toll of every ship that comes to the said haven at Sandwich, whatever it be and wherever it comes from. And if there is anything in the open sea outside the haven, the rights of the monks shall extend as far as the utmost limit to which the sea recedes and the length of a man in addition who holds a pole in his hand and stretches it as far as he can reach into the sea, and half of all that is found on this side of the "middle sea" and brought to Sandwich, whether it be clothes or nets or weapons, iron, gold, or silver, shall fall to the monks and the other half shall be left to those who find it. And if any other document is henceforth produced which was made heretofore and which in any kind of way seems to gainsay what is here

---

SOURCE: A. J. Robertson (ed.), *Anglo-Saxon Charters* (London: Cambridge University Press, 1956), 158–61.

established, that document shall be cast to mice to gnaw or into the fire to be burned, and he who produces it, whatever his rank, shall be regarded as the sweepings of ashes and confounded with the most ignominious shame and with one accord shunned by all the men who are nearby. And what is firmly established by this charter shall be enforced for all time henceforth, and with the authority of Almighty God and with mine and with the ratification of all my thegns it shall be established with lasting justice, steadfast and immovable forever against all attempts to gainsay it. And if there is anyone who attempts with swollen pride either to violate or to diminish what we here establish, he shall find himself accursed by God and by all his saints, unless before his death he make amends with due repentance for the sin that he has unrighteously committed.

This charter was written in the year 1023 after the birth of our Lord, the Saviour Christ, with the consent of the men whose names are recorded hereafter . . .

SECTION F

Alfred and the Struggle with the Scandinavians

21. Divided We Fall

Toward the end of the eighth century England was subjected to pillage and conquest on the part of warriors from the Scandinavian lands, whose seamanship, valor and lack of Christian scruples (they being still pagan) seemed likely, in the course of the following generations, to subdue completely the English peoples, and portions of Scotland and Ireland to boot. In 797 the English scholar Alcuin, then living at Charlemagne's court, sought to rally the people of Kent, warning them of their disunity in the face of imminent peril.

Let the nobles of the nation rule their lordships with the help of their councils, and preside over the people with justice, in their decisions loving the laws of their ancestors rather than money, which subverts the words of the righteous; and let them perform manfully, with one consent, what shall be for your benefit. Place over yourselves rulers famous for their

---

SOURCE: Albert S. Cook and Chauncey B. Tinker, *Select Translations from Old English Prose* (Boston: Ginn, 1908), 273–74.

nobleness, pious with the dignity of character, honorable with the beauty of righteousness, so that the divine mercy may vouchsafe to govern, preserve, and exalt your race.

A great danger threatens this island and the people living in it. A heathen people – and this was never known before – has accustomed itself to ravage our coasts with piratical depredations. Yet the peoples of England, her kingdoms and her kings, are at variance with one another. Hardly one of the ancient stock of our kings is left – I say it with tears – and the more uncertain their lineage the less is their courage. In like manner the teachers of truth have perished throughout the churches of Christ. Almost everyone follows the vanities of the world, and hates regular discipline, even the soldiers caring more for greed than for righteousness. Read Gildas, the wisest of the Britons, and you will see from what causes the ancestors of the Britons lost their kingdom and country; then if you will consider yourselves, you will find almost the same things there.

Fear for yourselves the declaration of the Truth Himself: "If a kingdom be divided against itself, that kingdom cannot stand."[38] See what division there is among the peoples and races of England. They are lacking in themselves because they do not keep peace and faith with one another.

Call back to yourselves, if you can make up your minds to it, your bishop Ethelheard,[39] a wise and venerable man; and according to his counsel improve the condition of your kingdom, and amend in conduct what is displeasing to God.

22. Alfred's Victory at Edington, 878

The turning point in the long drawn out struggle came in 878, when King Alfred defeated the Danes at Edington, in present-day Wiltshire, after Wessex had been virtually conquered. The engagement proved to be decisive in that the continuance of the West Saxon regime was assured.

878 – In this year in midwinter after Twelfth Night[40] the enemy army came stealthily to Chippenham, and occupied the land of the West Saxons

---

SOURCE: Dorothy Whitelock (ed.), *The Anglo-Saxon Chronicle* (London: Eyre and Spottiswoode, 1961; New Brunswick: Rutgers University Press), 49–50.

---

[38] Mark 3: 24.
[39] Archbishop of Canterbury, 791–805. Because of the hostility of the Kentishmen he was a refugee at the Mercian court.
[40] The evening of Epiphany, January 6.

and settled there, and drove a great part of the people across the sea, and conquered most of the others; and the people submitted to them, except King Alfred. He journeyed in difficulties through the woods and fen-fastnesses[41] with a small force. . . .

And afterwards at Easter, King Alfred with a small force made a stronghold at Athelney, and he and the section of the people of Somerset which was nearest to it proceeded to fight from that stronghold against the enemy. Then in the seventh week after Easter he rode to "Egbert's stone" east of Selwood and there came to meet him all the people of Somerset and of Wiltshire and of that part of Hampshire which was on this side of the sea, and they rejoiced to see him. And then after one night he went from that encampment to Iley, and after another night to Edington, and there fought against the whole army and put it to flight, and pursued it as far as the fortress, and stayed there a fortnight. And then the enemy gave him preliminary hostages and great oaths that they would leave his kingdom, and promised also that their king should receive baptism, and they kept their promise. Three weeks later King Guthrum with 30 of the men who were the most important in the army came (to him) at Aller, which is near Athelney, and the king stood sponsor to him at his baptism there; and the unbinding of the chrism[42] took place at Wedmore. And he was twelve days with the king, and he honored him and his companions greatly with gifts.

23. The Treaty between Alfred and Guthrum, 886–890

In 886 Alfred gained control of London, and forced the Danish leader Guthrum to sign a peace by which Alfred's dominion over the south-western half of England was acknowledged. The region to the north and east of the boundary described in the treaty was later known as the Danelaw, for there the Scandinavian authority was recognized.

Prologue. This is the peace which King Alfred and King Guthrum and the councillors of all the English race and all the people which is in East Anglia have all agreed on and confirmed with oaths, for themselves, and for their subjects, both for the living and those yet unborn, who care to have God's grace or ours.

SOURCE: F. Liebermann, *Die Gesetze der Angelsachsen* (Halle: Max Niemeyer, 1903–1916), *I*, 126, 128, in Dorothy Whitelock (ed.), *English Historical Documents, c. 500–1042,* David C. Douglas (gen. ed.), (London: Eyre and Spottiswoode, 1955; New York: Oxford University Press), 380–81.

---

[41] Low, marshy land.

[42] A garment used in baptism.

1. First concerning our boundaries: up the Thames, and then up the Lea, and along the Lea to its source, then in a straight line to Bedford, then up the Ouse to the Watling Street.[43]

2. This is next, if a man is slain, all of us estimate Englishman and Dane at the same amount, at eight half-marks[44] of refined gold, except the *ceorl* who occupies rented land, and their [the Danes'] freedmen; these also are estimated at the same amount, both at 200 shillings.

3. And if anyone accuses a king's thegn of manslaughter, if he dares to clear himself by oath, he is to do it with 12 king's thegns; if anyone accuses a man who is less powerful than a king's thegn, he is to clear himself with 12 of his equals and with one king's thegn – and so in every suit which involves more than four mancuses[45] – and if he dare not [clear himself], he is to pay threefold compensation, according as it is valued.

4. And that each man is to know his warrantor at [the purchase of] men or horses or oxen.

5. And we all agreed on the day when the oaths were sworn, that no slaves nor freemen might go without permission into the army of the Danes, any more than any of theirs to us. But if it happens that from necessity any one of them wishes to have traffic with us, or we with them, for cattle or goods, it is to be permitted on condition that hostages shall be given as a pledge of peace and as evidence so that one may know no fraud is intended.

24. Alfred: Patron of Religion and Learning

Alfred the Great (871–899) gained renown not only as a military commander and civil administrator, but also as a patron of learning. He brought out several translations of Latin works. In one of them, a version of Pope Gregory's *Pastoral Care,* Alfred revealed in his preface a concern over the decline of scholarship in his age, and presented a program to amend the situation.

King Alfred sends greetings to [blank][46] in loving and friendly words. And I would have you informed that it has very often come into my

---

SOURCE: Dorothy Whitelock (ed.), *English Historical Documents, c. 500–1042,* David C. Douglas (gen. ed.), (London: Eyre and Spottiswoode, 1955; New York: Oxford University Press), 818–19.

---

[43] The old Roman road to Chester.

[44] An old English money of account (not a coin), valued at 6s 8d, or half the value of the mark.

[45] The mancus, a money of account, was valued at thirty pence.

[46] Left blank so that the name of the recipient could be inserted.

mind what wise men there were in former times throughout England, both of spiritual and lay orders; and how happy times then were throughout England; and how the kings who had rule over the people were obedient to God and His messengers; and how they both upheld peace and morals and authority at home, and also extended their territory abroad; and how they prospered both in warfare and in wisdom; and also how zealous the spiritual orders were both about teaching and learning and all the services which they should do for God; and how foreigners came hither to this land in search of knowledge and instruction, and we now would have to get them from abroad, if we were to have them. So completely had learning decayed in England that there were very few men on this side the Humber who could apprehend their services in English or even translate a letter from Latin into English, and I think that there were not many beyond the Humber. There were so few of them that I cannot even recollect a single one south of the Thames when I succeeded to the kingdom. Thanks be to God Almighty that we now have any provision of teachers. And therefore I charge you to do, as I believe you are willing, detach yourself as often as you can from the affairs of this world, to the end that you may apply that wisdom which God has granted you wherever you may be able to apply it. Remember what temporal punishments[47] came upon us, when we neither loved wisdom ourselves nor allowed it to other men; we possessed only the name of Christians, and very few possessed the virtues.

When I remembered all this, I also remembered how, before everything was ravaged and burnt, the churches throughout all England stood filled with treasures and books, and likewise there was a great multitude of the servants of God. And they had very little benefit from those books, for they could not understand anything in them, because they were not written in their own language. As if they had said: "Our forefathers who formerly held these monasteries loved wisdom, and through it they obtained wealth and left it to us. Their track can still be seen, but we cannot follow it up, and we have now lost both the wealth and wisdom, because we were unwilling to incline our mind to that track."

When I remembered all this, I wondered exceedingly at those good and wise men who were in former times throughout England, and had fully studied all those books, that they would not turn any part of them into their own language. But then at once I answered myself, and said: "They did not think that men would ever become so careless and learning so decayed; they abstained intentionally, wishing that here in the land there should be the greater wisdoms, the more languages we knew."

Then I remembered also how the divine law was first composed in

[47] The reference is to the ravages of the Norsemen.

the Hebrew language, and afterwards, when the Greeks learnt it, they turned it all into their own language, and also all other books. And the Romans likewise, when they had learnt them, turned them all through learned interpreters into their own language. And also all other Christian nations turned some part of them into their own language. Therefore it seems better to me, if it seems so to you, that we also should turn into the language that we can all understand some books, which may be most necessary for all men to know; and bring it to pass, as we can very easily with God's help, if we have the peace, that all the youth now in England, born of free men who have the means that they can apply to it, may be devoted to learning as long as they cannot be of use in any other employment, until such time as they can read well what is written in English. One may then teach further in the Latin language those whom one wishes to teach further and to bring to holy orders.

When I remembered how the knowledge of the Latin language had previously decayed throughout England, and yet many could read things written in English, I began in the midst of the other various and manifold cares of this kingdom to turn into English the book which is called in Latin *Pastoralis* and in English "Shepherdbook," sometimes word for word, sometimes by a paraphrase; as I had learnt it from my Archbishop Plegmund, and my Bishop Asser, and my priest Grimbald and my priest John. When I had learnt it, I turned it into English according as I understood it and as I could render it most intelligibly; and I will send one to every see in my kingdom; and in each will be a book-marker [?] worth 50 mancuses.

And I command in God's name, that no one take that book-marker from the book, nor the book from the church. It is unknown how long there may be such learned bishops, as now, thanks be to God, are almost everywhere; therefore I would like them always to be at that place, unless the bishop wish to have it with him, or it is anywhere on loan, or someone is copying it. . . .

SECTION G

The Supremacy of Wessex and the Conquest by Canute

25. The Battle of Brunanburh, 937

Following Alfred's death in 899, his sons and grandsons strove to reconquer the Danelaw. By 918 all the Norsemen south of the Humber had acknowledged English rule, and in 927 King Athelstan carried English supremacy a step further by driving out the Norwegian King of York. Ten years later this extension of the power of Wessex was confirmed at the Battle of Brunanburh, where Athelstan and his brother defeated a combined force led by the Norse ruler of Dublin, the King of Scots and the King of Strathclyde. This marked the climax of the military achievement of the House of Wessex, and an unprecedented union among the British peoples. The battle was commemorated in the following poem, to be found in the Anglo-Saxon Chronicle; from its lines (as translated by Tennyson) we may sense something of the urgency, violence and magnitude of the accomplishment.

I

Athelstan King,
Lord among Earls,
Bracelet-bestower and
Baron of Barons,
He with his brother,
Edmund Atheling,
Gaining a lifelong
Glory in battle,
Slew with the sword-edge
There by Brunanburh,
Brake the shield-wall,
Hewed the linden-wood,[48]
Hacked the battle-shield,
Sons of Edward with hammered brands.

SOURCE: Albert S. Cook and Chauncey B. Tinker, *Select Translations from Old English Poetry* (Boston: Ginn, 1902), 26–30.

---

[48] I.e., shields of wood.

II

Theirs was a greatness
Got from their grandsires –
Theirs that so often in
Strife with their enemies
Struck for their hoards and their hearths and their homes.

III

Bowed the spoiler,
Bent the Scotsman,
Fell the ship-crews
Doomed to the death.
All the field with blood of the fighters
Flowed, from when first the great
Sun-star of morning-tide,
Lamp of the Lord God
Lord everlasting,
Glode over earth till the glorious creature
Sank to his setting.

IV

There lay many a man
Marred by the javelin,
Men of the Northland
Shot over shield.
There was the Scotsman
Weary of war.

V

We the West-Saxons,
Long as the daylight
Lasted, in companies
Troubled the track of the host that we hated.
Grimly with swords that were sharp from the grindstone,
Fiercely we hacked at the flyers before us.

VI

Mighty the Mercian,
Hard was his hand-play,
Sparing not any of
Those that with Anlaf,[49]

[49] Or Olaf, King of Northern Ireland.

Warriors over the
Weltering waters
Borne in the bark's-bosom,
Drew to this island –
Doomed to the death.

VII

Five young kings put asleep by the sword-stroke,
Seven strong earls of the army of Anlaf
Fell on the war-field, numberless numbers,
Shipmen and Scotsmen.

VIII

Then the Norse leader,
Dire was his need of it,
Few were his following,
Fled to his war-ship;
Fleeted his vessel to sea with the king in it,
Saving his life on the fallow flood.

* * *

XII

Then with their nailed prows
Parted the Norsemen, a
Blood-reddened relic of
Javelins over
The jarring breaker, the deep-sea billow,
Shaping their way toward Dyflen[50] again,
Shamed in their souls.

XIII

Also the brethren,
King and Atheling,
Each in his glory,
Went to his own in his own West-Saxonland,
Glad of the war.

XIV

Many a carcase they left to be carrion,
Many a livid one, many a sallow-skin –

[50] Dublin.

Left for the white-tailed eagle to tear it, and
Left for the horny-nibbed[51] raven to rend it, and
Gave to the garbaging war-hawk to gorge it, and
That gray beast, the wolf of the weald.

XV

Never had huger
Slaughter of heroes
Slain by the sword-edge –
Such as old writers
Have writ of in histories –
Hapt in this isle, since
Up from the East hither
Saxon and Angle from
Over the broad billow
Broke into Britain with
Haughty war-workers who
Harried the Welshman, when
Earls that were lured by the
Hunger of glory gat
Hold of the land.

26. The Reign of Canute

The conquest of all England by the Scandinavians finally came in 1017, after the reign of the ineffectual Ethelred, at the hands of Canute, son of Swegen, King of Denmark. Canute later succeeded to the Danish throne and conquered Norway and its dependencies, thus establishing a remarkably extensive domain. With such a man on the English throne, the country loomed larger in European affairs than ever before. In 1027 he visited Rome, partly as an act of devotion and partly from more materialistic motives. Canute could be ruthless, but the following letter, written from Rome, reveals a concern for the welfare of his subjects which was reflected in a period of peace and prosperity.

Canute, King of all England, and of Denmark, Norway and part of Sweden, to Ethelnoth,[52] Metropolitan, and Aelfric, Archbishop of York,

---

SOURCE: Thomas Forester (trans.), *The Chronicle of Florence of Worcester* (London: G. Bell and Sons, 1854), 137–38.

---

[51] Beaked.
[52] Archbishop of Canterbury, 1020–1038.

and to all the bishops and prelates and to the whole nation of the English, both the nobles and the commons, greeting:

I notify to you that I have lately taken a journey to Rome, to pray for the forgiveness of my sins, and for the welfare of my dominions and the people under my rule. I had long since vowed this journey to God, but I have been hitherto prevented from accomplishing it by the affairs of my kingdom and other causes of impediment . . .

Be it known to all of you that, at the celebration of Easter, a great assembly of nobles was present with our lord, the Pope John, and Conrad, the Emperor:[53] that is to say, all the princes of the nations from Mount Garganus[54] to the neighboring sea. All these received me with honor and presented me with magnificent gifts; but more especially was I honored by the Emperor with various gifts and valuable presents, both in gold and silver vessels and in palls and very costly robes. I spoke with the Emperor himself and the lord Pope and the princes who were there in regard to the wants of my people, English as well as Danes, that there should be granted to them more equal justice and greater security in their journeys to Rome, and that they should not be hindered by so many barriers on the road, nor harassed by unjust tolls. The Emperor assented to my demands, as well as King Rodolph,[55] in whose dominions these barriers chiefly stand; and all the princes made edicts that my people, the merchants as well as those who go to pay their devotions, shall pass to and fro in their journeys to Rome in peace, and under the security of just laws, free from all molestation by the guards of barriers or the receivers of tolls. I made further complaint to my lord the Pope, and expressed my high displeasure, that my archbishops are sorely aggrieved by the demand of immense sums of money when, according to custom, they resort to the apostolical see to obtain the pallium;[56] and it is decreed that it should no longer be done. All things, therefore, which I requested for the good of my people from my lord the Pope and the Emperor and King Rodolph and the other princes through whose territories our road to Rome lies, they have most freely granted, and even ratified their concessions by oath . . . Wherefore I return most humble thanks to Almighty God for my having successfully accomplished all that I had desired, as I had resolved in my mind, and having satisfied my wishes to the fullest extent.

Be it known therefore to all of you that I have humbly vowed to the Almighty God himself henceforward to amend my life in all respects, and to rule the kingdoms and the people subject to me with justice and clemency, giving equitable judgments in all matters; and if, through the

53 Conrad II, who was crowned on this occasion.
54 The most easterly spur of the Neapolitan Apennines.
55 Rodolph III, last King of Burgundy.
56 A vestment, bestowed by the Pope on prelates as a mark of their rank.

intemperance of youth or negligence, I have hitherto exceeded the bounds of justice in any of my acts, I intend by God's aid to make an entire change for the better. I therefore adjure and command my counsellors to whom I have entrusted the affairs of my kingdom, that henceforth they neither themselves commit, nor suffer to prevail, any sort of injustice throughout my dominions, either from fear of me or from favor to any powerful person. I also command all sheriffs and magistrates throughout my whole kingdom, as they tender my regard and their own safety, that they use no unjust violence to any man, rich or poor, but that all, high or low, rich or poor, shall enjoy alike impartial law; from which they are never to deviate, either on account of royal favor, respect of person in the great, or for the sake of amassing money wrongfully, for I have no need to accumulate wealth by iniquitous exactions. . . .

HARO LD:REX:INTERFEC
TUS:EST

Part 2

THE FEUDAL MONARCHY

1066-1272

SECTION A

The Norman Conquest

❖❖

27. The Invasion by William of Normandy, 1066

For a near-contemporary account of the conquest of England by William of Normandy, and of the events leading up to it, we may turn to William of Malmesbury (d. 1143?), who wrote, among other things, *Gesta Regum Anglorum* (Deeds of the Kings of the English), from which the following is taken:

King Edward, declining into years, as he had no children himself and saw the sons of Godwin[1] growing in power, sent messengers asking the King of Hungary to send over Edward,[2] the son of his brother Edmund, with all his family; intending, as he declared, that either he or his sons should succeed to the hereditary kingdom of England, and that his own want of issue should be supplied by that of his kindred. In consequence Edward came, but died almost immediately at St. Paul's, in London. He was neither valiant, nor a man of abilities. He left three surviving children: that is to say, Edgar, who after the death of Harold was by some elected king, and who after many revolutions of fortune is now living retired in the country in extreme old age; Christina, who grew old at

SOURCE: William of Malmesbury, *Gesta Regum Anglorum*, William Stubbs (ed.), (London: Rolls Series, 1887–89), *I*, 278–79; *II*, 297–98, 299–300, 302–03, in Edward P. Cheyney, *Readings in English History Drawn from the Original Sources* (Boston: Ginn, 1935), 90–95.

[1] Godwin, Earl of Wessex (d. 1053); his son, Harold, became King of England in 1065.
[2] Edward the Atheling, or heir apparent, who was in exile.

Romsey in the habit of a nun; and Margaret, whom Malcolm, King of the Scots, espoused. . . .

The King, in consequence of the death of his relation, losing his first hope of support, gave the succession of England to William, Duke of Normandy. He was well worthy of such a gift, being a young man of superior mind, who had raised himself to the highest eminence by his unwearied exertions; moreover, he was his nearest relation to consanguinity, as he was the son of Robert, the son of Richard the Second, whom we have repeatedly mentioned as the brother of Emma, Edward's mother. Some affirm that Harold himself was sent into Normany for this purpose; others, who knew Harold's more secret intentions, say that, being driven thither against his will by the violence of the wind, he imagined this device in order to extricate himself. This, as it appears nearest the truth, I will relate.

Harold, being at his country seat at Bosham, went for recreation on board a fishing boat, and for the purpose of prolonging his sport put out to sea; when, a sudden tempest arising, he was driven with his companions on the coast of Ponthieu. The people of that district, as was their native custom, immediately assembled from all quarters; and Harold's company, unarmed and few in numbers, were, as it easily might be, quickly overpowered by an armed multitude and bound hand and foot. Harold, craftily meditating a remedy for this mischance, sent a person, whom he had allured by very great promises, to William to say that Harold had been sent into Normandy by the King for the purpose of expressly confirming in person the message which had been imperfectly delivered by people of less authority, but that he was detained in fetters by Guy, Count of Ponthieu, and could not execute his embassy; that it was the barbarous and inveterate custom of the country that such as had escaped destruction at sea should meet with perils on shore; that it well became a man of William's dignity not to let this pass unpunished. If he should suffer those who appealed to his protection to be laden with chains, it would detract somewhat from his own greatness; and if his captivity must be terminated by money, he would gladly give it to Duke William, but not to the contemptible Guy. By these means Harold was liberated at William's command, and conducted to Normandy by Guy in person. The Duke entertained him with much respect, both in banqueting and vesture. . . .

When King Edward had yielded to fate, England, fluctuating with doubtful favor, was uncertain to which ruler she should commit herself — to Harold, William, or Edgar; for the King had recommended him also to the nobility, as the nearest to the sovereign in point of birth, concealing his better judgment because of the tenderness of his disposition. Wherefore, as I have said above, the English were distracted in their choice, although all of them openly wished well to Harold. He, indeed, once exalted by the diadem, thought nothing of the covenant[3] between himself and William, asserting that he was absolved from his oath because William's daugher, to whom he had been betrothed, had died before she was marriageable. For this man, though possessing numberless good qualities, is reported to have been careless about abstaining from perjury provided he could by any tricks elude the reasonings of men. Moreover, supposing that the threats of William would never be put into execution, because he was occupied in wars with neighboring princes, he, along with his subjects, had felt too much security.

Except for the fact that he had learned that the King of Norway was approaching, he would neither have condescended to collect troops nor to array them. William in the meantime began mildly to address him by messengers, to expostulate on the broken covenant, to mingle threats with entreaties, to warn him that ere a year expired he would claim his due by the sword, and that he would come to that place where Harold supposed he himself had the firmer footing. Harold again rejoined what I have related concerning the nuptials of his daughter, and added that he had been precipitate on the subject of the kingdom in having confirmed to him by oath another's right without the universal consent and edict of the general meeting and of the people; again, that a rash oath ought to be broken. . . .

At that time the prudence of William, seconded by the providence of God, already looked to England with hope, and that no rashness might stain the just cause he sent to the Pope – formerly Anselm, Bishop of Lucca, who had assumed the name of Alexander – alleging the justice of the war which he had undertaken with all the eloquence that he could.

Harold omitted to do this, either because he was confident by nature, or else distrusted his cause, or because he feared that his messengers would be obstructed by William and his partisans, who beset every port. The Pope, duly examining the pretensions of both parties, delivered a standard to William as an auspicious presage of the kingdom; on receiving which, the latter summoned an assembly of his nobles at Lillebourne for the purpose of ascertaining their sentiments on this undertaking. And when he had confirmed by splendid promises all who approved his design, he ordered all to prepare supplies of ships in proportion to the extent of their possessions. Thus they departed at that time, and in the month of August reassembled in a body at St. Valéry – for so that port is called by its new name. Collecting therefore ships from every quarter, they awaited the propitious gale which was to carry them to their destination. When this delayed blowing for several days, the common soldiers, as is generally

[3] The oath supposedly taken by Harold at Bayeux to support Duke William's interests.

the case, began to mutter in their tents that the man must be mad who wishes to subjugate a foreign country; that God was opposing them by withholding the wind; that William's father had planned a similar attempt, which was in like manner frustrated; that it was the fate of that family to aspire to things beyond their reach, and find God for their adversary. In consequence of these things, which were enough to enervate the force of the brave, being publicly noised abroad, the Duke held a council with his chiefs and ordered the body of St. Valéry to be brought forth and exposed to the open air for the purpose of imploring a wind. No delay now interposed, but the wished-for gale filled their sails. A joyful clamor then arising summoned every one to the ships. The Duke himself, first launching from the continent into the deep, awaited the rest at anchor, nearly in mid-channel. All then assembled around the crimson sails of the admiral's ship, and having first dined they arrived after a favorable passage at Hastings. As the Duke disembarked he slipped down, but turned the accident to his advantage, a soldier who stood near calling out to him, "You hold England, my lord, its future king." He then restrained his whole army from plundering, warning them that they should now abstain from what must hereafter be their own; and for fifteen successive days he remained so perfectly quiet that he seemed to think of nothing less than of war.

In the meantime Harold returned from the battle with the Norwegians, happy, in his own estimation, at having conquered, but not so in mine, as he had secured the victory by fratricide. When the news of the arrival of the Normans reached him, reeking as he was from battle, he proceeded to Hastings, though accompanied by very small forces. . . .

The courageous leaders mutually prepared for battle, each according to his national custom. The English, as we have heard, passed the night without sleep, in drinking and singing, and in the morning proceeded without delay towards the enemy. All were on foot, armed with battle-axes. Covering themselves in front by the junction of their shields, they formed an impenetrable body which would have secured their safety that day had not the Normans by a feigned flight induced them to open their ranks, which till that time, according to their custom, were closely compacted. The King himself, on foot, stood with his brother near the standard, in order that, while all shared equal danger, none might think of retreating. This standard William sent after the victory to the Pope. It was sumptuously embroidered of gold and precious stones, in the form of a man fighting.

On the other hand, the Normans passed the whole night in confessing their sins, and received the sacrament in the morning. Their infantry, with bows and arrows, formed the vanguard, while their cavalry, divided into wings, were thrown back. The Duke with serene countenance, declaring aloud that God would favor his as being the righteous side, called for his arms; and presently, when through the hurry of his attendants he had put on his hauberk the hindpart before, he corrected the mistake with a laugh, saying, "My dukedom shall be turned into a kingdom." Then, beginning the song of Roland, that the warlike example of that man might stimulate the soldiers, and calling on God for assistance, the battle commenced on both sides.

They fought with ardor, neither giving ground for great part of the day. Finding this, William gave a signal to his party that by a feigned flight they should retreat. Through this device the close body of the English, opening for the purpose of cutting down the straggling enemy, brought upon itself swift destruction; for the Normans, facing about, attacked them thus disordered, and compelled them to flee. In this manner, deceived by a stratagem, they met an honorable death in avenging their country. Nor were they at all wanting to their own revenge, as by frequently making a stand they slaughtered their pursuers in heaps; for, getting possession of an eminence, they drove down the Normans, when roused with indignation and anxiously striving to gain the higher ground, into the valley beneath, where, easily hurling their javelins and rolling down stones on them as they stood below, they destroyed them to a man. Besides, by a short passage with which they were acquainted, avoiding a deep ditch, they trod under foot such a multitude of their enemies in that place that they made the hollow level with the plain by the heaps of carcasses. This vicissitude of first one party conquering and then the other prevailed as long as the life of Harold continued; but when he fell, from having his brain pierced with an arrow, the flight of the English ceased not until night.

28. The Devastation of the North

Though William of Normandy was victorious at Hastings and was crowned on Christmas Day, 1066, as King of England, it took several years to stamp out rebellion and disaffection. The northerners, in particular, aided by the King of Denmark, challenged William's authority. They were subdued in 1069 and, by way of punishment to the offenders and warning to William's other subjects, were treated very harshly. Between York and Durham the land was laid waste and men were driven from it, as noted by Simon, a monk connected with the Abbey of Durham:

SOURCE: Symeon of Durham, *Historia Regum*, Thomas Arnold (ed.), (London: Rolls Series, 1882–85), *II*, 188, in Edward P. Cheyney, *Readings in English History Drawn from the Original Sources* (Boston: Ginn, 1935), 101.

The Normans had devastated Northumbria and certain other counties of England in the preceding year; during this [1069] and the following year through almost all England, especially Northumbria and the counties lying near it, so great a famine arose that, since necessity forced them, they ate the flesh of human beings, horses, dogs and cats, and whatever usage shrinks from; so severe was it that some sold themselves into perpetual slavery, provided only they might in any way sustain their wretched lives; others started to go into exile, but falling on the journey lost their lives. It was horrible to see the dead bodies decaying in the houses, in the open spaces, and on the streets. The mass decaying with horrible stench, swarmed with worms. Nor was any one left to bury the dead, for all were wiped out either by sword or famine, or had departed from their homes on account of hunger. In the meantime the land was destitute of cultivators, and a broad wilderness existed for nine years. Between York and Durham nowhere was there an inhabited village, while the dens of wild beasts and robbers caused terror to travelers.

SECTION B

The Feudal Relationship

29. Homage and Fealty

Fundamental to the feudal relationship was the acknowledgment by the vassal that he would become his lord's man (in French, *homme*, from which comes the word "homage") and be faithful to him. Thus enfeoffments (feudal grants) of land were accompanied by formal acts of homage and fealty. The thirteenth century jurist, Bracton, describes these obligations in his important work, *De Legibus et Consuetudinibus Angliae* (Concerning the Laws and Customs of England), the earliest comprehensive treatment of the subject.

Be it known that he who should do his homage, in view of the reverence which he owes his lord, should wait upon his lord wherever he may be found in the realm or elsewhere, if he can conveniently be waited

SOURCE: Bracton, *De Legibus et Consuetudinibus Angliae,* Travers Twiss (ed.), (London: Rolls Series, 1878–83), *I,* 632, in A. E. Bland, *The Normans in England, 1066–1154* (London: G. Bell and Sons, 1914), 26–27.

upon; and the lord is not bound to seek out his tenant. And he should do homage to him in this wise: the tenant should first put both his hands between both hands of his lord, which signifies on the part of the lord protection, defence and warranty, and on the part of the tenant reverence and subjection, and he should say these words: "I become your man for the tenement which I hold of you and ought to hold, and I will bear faith to you of life and limb and earthly honor, and I will bear faith to to you against all men, saving the faith due to the lord the King and his heirs." And straightway afterwards he shall make the oath of fealty to his lord in this wise: "This you hear, my lord N., that I will bear faith to you of life and limb, body and chattels and earthly honor, so help me God and these holy relics." And some add in the oath, and well so, that faithfully and without diminution, contradiction or hindrance or unjust delay, the tenant will do his service to his lord and his heirs at the stated terms.

And homage ought not to be done in private, but in a public and common place before many persons in the county, hundred or court, so that if by chance the tenant through malice should wish to deny the homage, the lord could the more easily have proof of homage done and service acknowledged.

30. Enfeoffment

This document, proclaiming enfeoffment and the accompanying feudal obligations, refers to the reign of William I. It has been pointed out that the careful description of Peter's obligations "suggests that knight's tenure has as yet not been clearly defined."[4] It will be noted that Peter becomes the vassal of an ecclesiastic: most of the lands of the Church were held by military service.

Be it known to all of you that Peter, a knight of King William, will become the feudal man of St. Edmund and of Baldwin the abbot, by performing the ceremony of homage. He will do this by permission of the King and with the consent of the monks, and in return for the service which will here be stated, saving always the fealty which he owes to the King, the fief having been freely received except for the six royal forfeitures. Peter promises that he will serve on behalf of the abbot within

---

SOURCE: David C. Douglas and George W. Greenaway (eds.), *English Historical Documents, 1042–1189,* David C. Douglas (gen. ed.), London: Eyre and Spottiswoode, 1953; New York: Oxford University Press), 896–97.

---

[4] David C. Douglas and George W. Greenaway (eds.), *English Historical Documents, 1042–1189* (New York, Oxford, 1953), 896*n.*

the kingdom with 3 or 4 knights at their own expense if he has been previously summoned by the King and the Abbot to take part in the earlier or later levies of the King's host. If he is bidden to plead on the Abbot's behalf at any place within the kingdom, they shall likewise bear their own expense. But if the Abbot shall take him anywhere else, then the expense of his service shall be borne by the Abbot. Besides this, he shall equip a knight for service without or within the kingdom where and when the Abbot shall require to have this knight as his own retainer . . .

31. A Summons to the Feudal Host, c. 1072

This summons, which probably dates from about 1072, shows how the King called upon those who held by knight service to form his feudal army, or host. It is important in showing how soon after the Norman Conquest such arrangements came into being.

William, King of the English, to Aethelwig, Abbot of Evesham, greeting. I order you to summon all those who are subject to your administration and jurisdiction that they bring before me at Clarendon on the Octave of Pentecost[5] all the knights they owe me duly equipped. You, also, on that day, shall come to me, and bring with you fully equipped those 5 knights which you owe me in respect of your abbacy. Witness Eudo the steward. At Winchester.

32. The Lord-Vassal Relationship

The relations between lords and vassals were supposed to be based on cooperation and mutual respect. But the position of the lord was predominant, as is illustrated by these excerpts from a twelfth-century statement of legal custom known as *Leges Henrici Primi* (Laws of Henry I). The holding of land from several lords complicated feudal relations; in such cases one lord (the liege lord) had a prior claim to allegiance.

SOURCES: David C. Douglas and George W. Greenaway (eds.), *English Historical Documents, 1042–1189,* David C. Douglas (gen. ed.), London: Eyre and Spottiswoode, 1953; New York: Oxford University Press), 895.

Benjamin Thorpe, *Ancient Laws and Institutes* (London: Eyre and Spottiswoode, 1840), *I,* 552, 590, in Edward P. Cheyney, *Readings in English History Drawn from the Original Sources* (Boston: Ginn, 1935), 132.

---

[5] A week after the Festival of Pentecost.

It is allowable to anyone, without punishment, to support his lord, if anyone assails him, and to obey him in all legitimate ways, except in theft, murder, and in all things as are not conceded to anyone to do and are reckoned infamous by the laws.

The lord ought to do likewise equally with counsel and with aid; and he may come to his man's assistance in his vicissitudes in all ways, without forfeiture . . .

To every lord it is allowed to summon his man that he may be at right to him in his court; and even if he is resident at the most distant manor of that honor[6] from which he holds, he shall go to the place if his lord summons him. If his lord holds different fiefs, the man of one honor is not compelled by law to go to another plea, unless the cause belongs to the other to which his lord has summoned him.

If a man holds from several lords and honors, however much he holds from others, he owes most and will be subject for justice to him of whom he is the liegeman.

Every vassal owes to his lord fidelity concerning his life and members and earthly honor, and keeping of his counsel in what is honorable and useful, saving the faith of God and of the prince of the land. Theft, however, and treason and murder and whatever things are against the Lord and the Catholic faith are to be required of or performed by no one; but faith shall be held to all lords, saving the faith of the earlier and the more to the one of which he is the liege. And let permission be given him if any of his men seek another lord for himself.

SECTION C

The Central Government under the Normans

33. The Status of Kings

The theory that kings were men set apart from other men (and hence might exercise an authority beyond that conferred by feudal status),

---

SOURCE: H. Böhmer (ed.), *Monumenta Germanica Historia: Libelli de Lite Imperatorum et Pontificium* (Hanover: Hahn, 1891–97), *III*, 663, 665, in *English Historical Documents, 1042–1189*, David C. Douglas and George W. Greenaway (eds.), David C. Douglas (gen. ed.), (London: Eyre and Spottiswoode, 1953; New York: Oxford University Press), 676–77.

---

[6] Several manors held under one lord.

which can be descried in Anglo-Saxon times, was further developed after the Norman Conquest. This is evidenced in the following extract from a late eleventh or early twelfth-century tract, entitled "On the Consecration of Prelates and Kings," which also conveys the medieval idea of unity of church and state.

By divine authority and the institution of the holy fathers kings are consecrated in God's Church before the sacred altar and are anointed with holy oil and sacred benediction to exercise the ruling power over Christians, the Lord's people, "a chosen generation, an holy nation, a peculiar people," the holy Church of God. For what indeed is God's Church other than the congregation of faithful Christians dwelling together in one bond of faith, hope and charity within God's house? For the governance of the Church kings receive power at their coronation, to rule and confirm her in justice and judgment and to order her affairs according to the discipline of the Christian law. Thus they reign over the Church, which is the kingdom of God; they reign together with Christ in order to rule, protect and defend her. For to reign is to rule one's subjects well and to serve God in fear. To this same end are the bishops also instituted and consecrated with holy oil and sacred benediction, that they in their turn may govern the Church in accordance with the teaching handed down to them by God. For this cause the blessed Pope Gelasius wrote thus, "There are two orders by which this world is chiefly governed, the priestly authority and the royal power.". . . Notwithstanding, if the king has so great a dominion over the bodies of Christians, will he not also have dominion over God's sacred temple? . . . If this be so, it is plain that the king has dominion over those who obtain the priestly office. The king ought, therefore, not to be excluded from the governance of the Church, that is, the Christian people, because the kingdom would then be separated from the Church and destroyed. . . .

King and priest are both anointed with holy oil; they receive the same Holy Spirit of sanctification and the same virtue of benediction; they have a common name in Christ and a common interest, to whose merits they owe their name. . . .

King and priest have the same grace and certain privileges in common, but they have also their own peculiar and diverse offices. For although they may appear to have the power to rule in common, yet this power is to be exercised in one way by the priest and in another way by the king, and grace is accorded to each for the fulfillment of his office. . . .

No one ought by right to take precedence of the king, on whom is bestowed so many and great blessings, who is consecrated with so many and great sacraments and dedicated to God, because no one receives greater or better blessings or is consecrated and dedicated to God with greater or higher sacraments, not even with the same or like sacraments, and in this respect the king is unique. Wherefore he is not to be considered a mere layman, for he is the Lord's anointed and by adoption and grace made like unto God; he is the supreme ruler, the chief shepherd, master, defender and instructor of holy Church, lord over his brethren and worthy to be worshipped by all men as chief "bishop" and supreme ruler. Nor is he to be spoken of as inferior to the bishop, because the latter consecrates him, since it often happens that superiors are consecrated by their inferiors, just as the pope is consecrated by the cardinals and the metropolitan by his suffragans. This is done because they are not the authors of the consecration, but the ministers. God is the author of the sacrament, the bishops merely officiate as his agents. . . .

34. The Oath of Salisbury, 1086

William I insisted that, despite the chain of loyalties produced by the process of subinfeudation, men owed their principal loyalty to the King rather than to their immediate lord. He thus contributed to the strengthening of the monarchical authority. This principle was invoked at a meeting of the court at Salisbury in 1086, at which William received homage and fealty from his barons and their more important vassals. The Anglo-Saxon Chronicle describes the event as follows.

1086 – In this year the king wore his crown and held his court at Winchester for Easter, and travelled so as to be at Westminster for Whitsuntide, and there dubbed his son, Henry, a knight. Then he travelled about so as to come to Salisbury at Lammas;[7] and there his councillors came to him, and all the people occupying land who were of any account all over England, no matter whose vassals they might be; and they all submitted to him and became his vassals, and swore oaths of allegiance to him, that they would be loyal to him against all other men. . . .

35. The Coronation Charter, 1100

Despite recognition of the extraordinary authority of the crown, kings sometimes found it necessary to give assurance that they would abide

---

SOURCES: Dorothy Whitelock (ed.), *The Anglo-Saxon Chronicle* (London: Eyre and Spottiswoode, 1961; New Brunswick: Rutgers University Press), 162.

Benjamin Thorpe, *Ancient Laws and Institutes* (London: Eyre and Spottiswoode, 1840), *I*, 497–500, in Edward P. Cheyney, *Readings in English History Drawn from the Original Sources* (Boston: Ginn, 1935), 121–23.

---

[7] The wheat-harvest festival, August 1.

by customary feudal arrangements and make other concessions to secure the support and loyalty of their subjects. Thus Henry I, who gained the throne through neither primogeniture nor election, issued a charter of liberties known as the "Coronation Charter," in which he acknowledged limitations to his powers.

In the year of the incarnation of the Lord 1101, Henry, son of King William, after the death of his brother William, by the grace of God King of the English, to all faithful barons and men, . . . greeting.

1. Know that by the mercy of God, and by the common counsel of the barons of the whole kingdom of England, I have been crowned king of the same kingdom; and because the kingdom has been oppressed by unjust exactions, I, from regard to God, and from the love which I have toward you, in the first place make the holy church of God free, so that I will neither sell nor place at rent, nor, when archbishop, or bishop, or abbot is dead, will I take anything from the domain of the church, or from its men, until a successor is installed into it. And all the evil customs by which the realm of England was unjustly oppressed will I take away, which evil customs I partly set down here.

2. If any of my barons, or earls, or others who hold from me shall have died, his heir shall not redeem his land as he did in the time of my brother, but shall relieve it by a just and legitimate relief. Similarly also the men of my barons shall relieve their lands from their lords by a just and legitimate relief.

3. And if any of the barons or other men of mine wishes to give his daughter in marriage, or his sister or niece or relation, he must speak with me about it, but I will neither take anything from him for his permission, nor forbid him to give her in marriage, unless he should wish to join her to my enemy. And if when a baron or other man of mine is dead, a daughter remains as his heir, I will give her in marriage according to the judgment of my barons, along with her lands. And if when a man is dead his wife remains and is without children, she shall have her dowry and right of marriage, and I will not give her to a husband except according to her will.

4. And if a wife has survived with children, she shall have her dowry and right of marriage, so long as she shall have kept her body legitimately, and I will not give her in marriage, except according to her will. And the guardian of the land and children shall be either the wife or another one of the relatives, as shall seem to be most just. And I require that my barons should deal similarly with the sons and daughters or wives of their men. . . .

7. And if any one of my barons or men shall become feeble, however he himself shall give or arrange to give his money, I grant that it shall be

so given. Moreover, if he himself, prevented by arms or by weakness, shall not have bestowed his money, or arranged to bestow it, his wife or his children or his parents and his legitimate men shall divide it for his soul, as to them shall seem best.

8. If any of my barons or men shall have committed an offense, he shall not give security to the extent of forfeiture of his money, as he did in the time of my father, or of my brother, but according to the measure of the offense so shall he pay, as he would have paid from the time of my father backward, in the time of my other predecessors; so that if he shall have been convicted of treachery or of crime, he shall pay as is just.

9. All murders, moreover, before that day in which I was crowned king, I pardon; and those which shall be done henceforth shall be punished justly according to the law of King Edward.

10. The forests, by the common agreement of my barons, I have retained in my own hand, as my father held them.

11. To those knights who hold their land by the cuirass,[8] I yield of my own gift the lands of their demesne plows free from all payments and from all labor, so that as they have thus been favored by such a great alleviation, so they may readily provide themselves with horses and arms for my service and for the defense of the kingdom.

12. A firm peace in my whole kingdom I establish and require to be kept from henceforth.

13. The law of King Edward I give to you again, with those changes with which my father changed it by the counsel of his barons.

14. If any one has taken anything from my possessions since the death of King William, my brother, or from the possessions of any one, let the whole be immediately returned without alteration; and if any one shall have retained anything thence, he upon whom it is found shall pay it heavily to me. . . .

36. Inquest for Domesday Book, 1086

William I, as well as his successors, occasionally sent commissioners into various counties to conduct business there. In 1086 a particularly important assignment was given to such commissioners: William sought information as to the resources of the country in order that he might levy new and heavier tax assessments. The result was the com-

---

SOURCE: William Stubbs, *Select Charters and Other Illustrations of English Constitutional History* (Oxford: The Clarendon Press, 1895), 86, in A. E. Bland, *The Normans in England, 1066–1154* (London: G. Bell and Sons, 1914), 24.

---

[8] I.e., by military service; equipped with a hauberk of mail.

pilation of a record known as Domesday Book, which is unparalleled among European countries of the Middle Ages. The commissioners were instructed as follows:

Here below is written the inquest upon the lands, in what manner the King's barons make enquiry, to wit, by the oath of the sheriff of the shire, and of all the barons and their Frenchmen, and of the whole hundred, of the priest, the reeve, and six villeins of each town. Then how the manor is named; who held it in the time of King Edward; who holds it now; how many hides; how many ploughs on the demesne, and how many men; how many villeins; how many cotters;[9] how many serfs; how many freemen; how many socmen;[10] how much wood; how much meadow; how many pastures; how many mills; how many fishponds; how much has been added or taken away; how much it was worth altogether; and how much now; how much each freeman or socman there had or has. All this for three periods: to wit, in the time of King Edward; and when King William granted it; and as it is now; and if more can be had therefrom than is had.

37. The Exchequer

The financial position and pretensions of the English monarchs were greatly furthered, probably in the reign of Henry II, by the establishment of a department of government called the Exchequer. Richard fitz Nigel, Bishop of London and treasurer of the Exchequer from about 1160 to 1190, has described this institution in a work called the *Dialogue of the Exchequer,* from which the following is taken:

What the exchequer is, and what is the reason for its name.

DISCIPLE. What is the Exchequer?

MASTER. The Exchequer is a quadrangular board about ten feet in length and five in breadth placed before those who sit around it in the manner of a table. Running round it there is a raised edge about the

SOURCE: David C. Douglas and George W. Greenaway (eds.), *English Historical Documents, 1042–1189,* David C. Douglas (gen. ed.), (London: Eyre and Spottiswoods, 1953; New York: Oxford University Press), 493–94. Based on edition of the *Dialogue* by A. Hughes, C. G. Crump and C. Johnson (Oxford: The Clarendon Press, 1902).

[9] Villeins of low grade.

[10] Socage tenants, required to render specific services of various kinds.

height of four fingers, lest anything placed upon it should fall off. Above the board is placed a cloth purchased at the Easter term – not an ordinary cloth, but a black one marked with stripes which are the space of a foot or a hand's breadth distant from each other. Within the spaces counters are placed according to their value, and of this we shall speak elsewhere. But although such a board is called an "exchequer," nevertheless this name is applied also to the court which sits with the Exchequer so that if anyone obtains anything through its judgment, or anything is determined by common counsel, this is said to have been done "at the Exchequer," of such-and-such a year. Thus as men say today "at the Exchequer," so they formerly said "at the tallies."

DISCIPLE. What is the reason for this name?

MASTER. No truer one at present occurs to me than that it has a shape similar to that of a chess-board.

DISCIPLE. Would the ancients in their wisdom have so named it for its shape alone? For it might for a similar reason have been called a draught-board.

MASTER. I was right to call thee conscientious. There is another but less obvious reason. For just as in a game of chess there are certain classes of pieces and they move or stand still by definite rules or within certain limits, some being first in dignity and others in place, so in this, some preside while others sit in virtue of their office and none is free to depart from the established rules, as will be manifest from what follows. Again, as in chess the battle is joined by the kings, so in this it is chiefly between two men that the conflict takes place and the battle is waged, namely, the treasurer and the sheriff who sits there to render his account, while the others sit by like umpires to watch and judge the proceedings.

DISCIPLE. It is then the treasurer who receives the account, although there are many others present who are of higher rank?

MASTER. That the treasurer should receive the account from the sheriff is manifest from the fact that the said account is required from him whenever it shall please the king: for what he had not received could hardly be demanded of him. Nevertheless, there are some who say that the treasurer and the chamberlains are only answerable for what is entered in the rolls as being "In the Treasury," and for this alone an account may be demanded of them. But the more correct view is that they should answer for all that is written in the roll. . . .

SECTION D

Legal and Judicial Developments

38. Presentment of Englishry

As far as the native English were concerned, the Norman conquest imposed an alien rule on their land. To counteract violence against the new ruling class, William I introduced a procedure known as presentment of Englishry: whenever a dead body was discovered, those of the neighborhood were called upon to prove that the slain man was English, or submit to a heavy fine. Richard fitz Nigel describes this procedure in his "Dialogue of the Exchequer."

Now in the primitive state of the kingdom, after the Conquest, the remnant of the conquered English secretly laid ambushes for the Normans whom they distrusted and hated, and far and wide, in woods and remote places, when opportunity presented itself, they slew them in secret. When, to avenge their deaths, the monarchs and their ministers had for some years taken violent measures against the English with various refinements of torture, and yet the latter had not altogether ceased their attacks, at length the following plan was devised, namely that the hundred in which a Norman was found slain in this way – no evidence being found as to the identity of the slayer – should be condemned to pay into the Treasury a large sum of tested silver; some indeed to the extent of thirty-six pounds, others, forty-four pounds, according to the different localities and the frequency of homicide. It is said that this was done so that the imposition of a general penalty might make it safe for wayfarers, and each man might hasten to punish so great a crime and to hand over to justice the man by whose fault such an enormous indemnity was imposed on the whole neighborhood. . . .

39. The Judicial Duel

William the Conqueror, though a stern ruler, did not seek to alter materially English governmental and legal institutions and practices.

SOURCE: David C. Douglas and George W. Greenaway (eds.), *English Historical Documents, 1042–1189,* David C. Douglas (gen. ed.), (London: Eyre and Spottiswoode, 1953; New York: Oxford University Press), 522–23. Based on edition of the *Dialogue* by A. Hughes, C. G. Crump and C. Johnson (Oxford, The Clarendon Press, 1902).

Only three pieces of legislation have been preserved from his reign. One of these introduced the method of judicial trial called wager of battle, or the judicial duel, well known in Normandy but not hitherto resorted to in England.

William by the grace of God King of the English, to all to whom this writing shall come greeting and friendship. We order and require this to be kept by the whole nation of England.

If an Englishman shall summon any Frenchman to battle for a theft or a homicide or any other matter for which battle ought to be waged or a plea made between the two men, he shall have full liberty to do this. And if the Englishman does not wish a battle, the Frenchman who is accused may defend himself by an oath against him, by his witnesses, according to the law of Normandy.

Likewise if a Frenchman shall summon an Englishman to battle concerning the same matters, the Englishman may with full liberty defend himself by battle, or by compurgation if that pleases him better. And if he is sickly and does not wish a battle, or is not competent, let him seek for himself a legal defender. If the Frenchman shall have been conquered, let him pay sixty shillings to the king. And if the Englishman does not wish to defend himself by battle, or by testimony, let him defend himself by the judgment of God.[11]

40. William I's Ordinance on Church Courts

Although William I sought to prevent the Church from encroaching on what he considered the necessary authority of the royal government, he encouraged the separation of lay and ecclesiastical authority in courts and councils. Particularly noteworthy is his edict by which matters pertaining to the church were not to be tried in the hundred courts, but before the bishops elsewhere, which contributed to friction between church and state in years to come.

SOURCES: Benjamin Thorpe, *Ancient Laws and Institutes* (London: Eyre and Spottiswoode, 1840), *I*, 488, in Edward P. Cheyney, *Readings in English History Drawn from the Original Sources* (Boston: Ginn, 1935), 105.

Benjamin Thorpe, *Ancient Laws and Institutes* (London: Eyre and Spottiswoode, 1840), *I*, 495–96, in Edward P. Cheyney, *Readings in English History Drawn from the Original Sources* (Boston: Ginn, 1935), 109–10.

---

[11] The ordeal.

William, by the grace of God King of the English, to R. Bainard and G. de Magneville and P. de Valoines and all my liege men of Essex, Hertfordshire, and Middlesex, greeting.

Know ye, and all my liege men resident in England, that I have by my common council and by the advice of the archbishops, bishops, abbots, and chief men of my realm, determined that the episcopal laws be amended, since they have not been kept properly nor according to the decrees of the sacred canons throughout the realm of England, even to my own times. Accordingly I command and charge you by royal authority that no bishop nor archdeacon do hereafter hold pleas concerning the episcopal laws in the hundred, nor bring to the judgment of secular men a cause which concerns the rule of souls. But whoever shall be impleaded for any cause or crime by the episcopal laws, let him come to the place which the bishop shall choose and name for the purpose, and there answer for his cause or crime, and not according to the hundred, but according to the canons and episcopal laws; and let him do right to God and his bishop.

But if any one, being lifted up with pride, refuse to come to the bishop's court, let him be summoned a first, second, and third time; if he does not then come to the judgment, let him be excommunicated; and if there is need of carrying this out, let the strength and justice of the king or of the sheriff be brought to bear. He who, summoned to the judgment of the bishop, refuses to come, shall answer to the bishop's law for each summons. This also I forbid and by my authority prohibit, that any sheriff or reeve or minister of the king or any layman should interfere in the laws that pertain to the bishop, or any layman should bring another to judgment without the justice of the bishop. Judgment, moreover, shall not be given in any place except in the bishop's see, or in such a place as the bishop shall have appointed for it.

41. Early Jury Procedure

In the history of English law, the reign of Henry II is of extraordinary importance. In the interest of public order, Henry issued in 1166 a decree called the Assize of Clarendon. Among its provisions was the extension of the inquest procedure to the investigation of certain crimes. Here we have the effective beginning of the indicting, or charging, jury, commonly known as the grand jury.

SOURCE: William Stubbs, *Select Charters and Other Illustrations of English Constitutional History* (Oxford: The Clarendon Press, 1895), 143–46, in Carl Stephenson and Frederick G. Marcham, *Sources of English Constitutional History* (New York: Harper and Row, 1937), 76–79.

Here begins the Assize of Clarendon made by King Henry, namely the second, with the assent of the archbishops, bishops, abbots, earls, and barons of all England.

1. In the first place the aforesaid King Henry, by the counsel of all his barons, has ordained that, for the preservation of peace and the enforcement of justice, inquiry shall be made in every county and in every hundred through twelve of the more lawful men of the hundred and through four of the more lawful men of each vill, on oath to tell the truth, whether in their hundred or in their vill there is any man accused or publicly known as a robber or murderer or thief, or any one who has been a receiver of robbers or murderers or thieves, since the lord King has been king. And let the justices make this investigation in their presence and the sheriffs in their presence.

2. And whoever is found by the oath of the aforesaid men to have been accused or publicly known as a robber or murderer or thief, or as a receiver of them since the lord King has been king, shall be seized; and he shall go to the ordeal of water and swear that, to the value of 5s, so far as he knows, he has not been a robber or murderer or thief, or a receiver of them, since the lord King has been king. . . .

14. The lord King also wills that those who make their law and are cleared by the law, if they are of very bad reputation, being publicly and shamefully denounced by the testimony of many lawful men, shall abjure the lands of the King, so that they shall cross the sea within eight days unless they are detained by the wind; then, with the first [favorable] wind that they have, they shall cross the sea and thenceforth not return to England, except at the mercy of the lord King; so that they shall there be outlaws and shall be seized as outlaws if they return . . .

SECTION E

The Relations of Church and State

42. William I and the Papacy

William I, though to some extent indebted to the Church for its support of his invasion, and in sympathy with increasing its spiritual authority, set restraints on the exercise of the papal power in his new kingdom. In 1076 he informed Pope Gregory VII, who sought to establish both the supremacy of the papacy within the Church, and of the Church over the secular state, that although willing to make customary payments he would not perform the act of fealty.

Your legate Hubert, most holy father, coming to me on your behalf, has admonished me to profess allegiance to you and your successors, and to think better regarding the money which my predecessors were wont to send to the Church of Rome. I have consented to the one but not to the other. I have not consented to pay fealty, nor will I now, because I never promised it, nor do I find that my predecessors ever paid it to your predecessors. The money has been negligently collected during the past three years when I was in France; but now that I have returned by God's mercy to my kingdom, I send you by the hands of the aforesaid legate what has already been collected, and the remainder shall be forwarded by the envoys of our trusty Archbishop Lanfranc when the opportunity for so doing shall occur. Pray for us and for the state of our realms, for we always loved your predecessors and it is our earnest desire above all things to love you most sincerely and to hear you most obediently.

Eadmer, a contemporary of William, who as a monk at Canterbury produced a chronicle of the events of the time (*Historia Novorum*), comments further on the King's limitations on ecclesiastical authority:

1. He would not then allow anyone settled in all his dominion to acknowledge as apostolic the Pontiff of the city of Rome, save at his own

---

SOURCES: J. A. Giles (ed.), *Lanfranci Opera* (Oxford: Parker, 1844), *I*, 32, in *English Historical Documents, 1042–1189*, David C. Douglas and George W. Greenaway (eds.), David C. Douglas (gen. ed.), (London: Eyre and Spottiswoode, 1953; New York: Oxford University Press), 647.

Eadmer of Canterbury, *Historia Novorum in Anglia*, Martin Rule (ed.), (London: Rolls Series, 1884), 10, in Edward P. Cheyney, *Readings in English History Drawn from the Original Sources* (Boston: Ginn, 1935), 110–11.

bidding; or by any means to receive any letter from him if it has not first been shown to himself.

2. The Primate also of his realm (I mean the Archbishop of Canterbury), presiding over a general council assembled of bishops, he did not permit to ordain or forbid anything save what had first been ordained by himself as agreeable to his own will.

3. He would not suffer that any, even of his bishops, should be allowed to implead publicly, or excommunicate or constrain by any penalty of ecclesiastical rigor, any of his barons or ministers accused of incest or adultery or any capital crime, save by his command.

43. The Investiture Controversy

By investiture is meant the conferring of an office or dignity. In England and elsewhere a long-drawn controversy developed as to whether the right of investiture in high ecclesiastical positions should be exercised by the temporal or the spiritual authorities. Since bishops and abbots were frequently great landholders, holding by military service, the crown felt impelled to control their appointment and to receive homage from them. The church repudiated this claim. This was the basis for the quarrel between Anselm, Archbishop of Canterbury, and Henry I, as described in these passages from the chronicler Eadmer. In the end a compromise was worked out, whereby the crown obtained the right to receive homage for lands, but gave up its claim to invest ecclesiastics with the symbols of their spiritual office, such as the bishop's crozier and ring.

A few days after his return, Anselm came to the King at Salisbury and was welcomed by him; he accepted the King's excuse for having assumed the royal dignity without waiting for the benediction of him whose right he knew it to be, and was thereupon required to do homage to the King according to the custom of his ancestors, and to receive the archbishopric from the King's hand. He answered that in no wise either would he or could he consent so to do, and when asked why, he immediately set forth in plain words what he had agreed to on these and certain other matters in the council at Rome, saying in conclusion, "If the lord the King will accept these terms, and accepting, observe them, there shall surely be a firm peace between us; but if not, I do not see that my remaining in England will be either useful or honest; especially as, if he has granted

---

SOURCE: Eadmer of Canterbury, *Historia Novorum in Anglia,* Martin Rule (ed.), (London: Rolls Series, 1884), 119–21, 131–34, 186, in A. E. Bland, *The Normans in England, 1066–1154* (London: G. Bell and Sons, 1914), 65–68, 71–72.

any bishoprics or abbacies, I must altogether reject communion both with him and with those who have accepted them. I have not returned to England to dwell there, unless the King will obey the Pope of Rome. Therefore I beg that the King will make what order he will, that I may know which way to turn."

The King, on hearing this, was gravely disturbed. It seemed to him a serious matter to lose the investitures of churches and the homage of prelates, but not less serious to suffer Anselm to leave the realm before he himself was fully established on the throne. On the one hand he thought he would be losing as it were half the realm, and on the other he feared that Anselm would go to his brother Robert,[12] who had by that time returned to Normandy from Jerusalem, and persuading him to submit to the apostolic see, which he knew to be a most easy thing to do, would make him king of England. A truce, therefore, from controversy on either side was asked for until Easter. . . . To this Anselm consented. . . .

Not long after . . . a friendly letter was sent by the King to him . . . asking him to come to the King, who wished the matter to be settled and had another plan. Hoping to hear that God of his grace had touched the King's heart, he went, as he was ordered, to Winchester. There the bishops and chief men of the realm were gathered together, and by their common assent Anselm agreed that . . . envoys should be sent by both parties to Rome to explain to the Roman pontiff face to face that either he must abandon his original decision, or submit to the expulsion of Anselm and his party from England and lose the submission of the whole realm and the profits which he was accustomed to derive yearly from the same. Two monks therefore were sent by Anselm, to wit, Baldwin of Bec and Alexander of Canterbury, not indeed to urge the Roman pontiff in any way to abate the rigor of justice on Anselm's behalf, but partly to bear testimony of the threats of the court which the Pope must straightway believe, and partly to bring back to Anselm a final decision from the apostolic see. To accomplish the same purpose the King sent three bishops, Gerard of Hereford, lately made Archbishop of York, Herbert of Thetford and Robert of Chester. . . .

The journey at length accomplished, the envoys reached Rome together, and announced the cause of their coming to the apostolic ears, each party presenting its own case, and humbly asked for the Pope's counsel to put an end to the quarrel. He heard their story and found no words in which to express his amazement. But when he was urgently pressed by the bishops to consult his own interests and mitigate the strictness of his predecessor's rigid decision, that peace might everywhere abound, he

[12] Robert, Duke of Normandy, the eldest son of William the Conqueror.

declared that he would not do it even to ransom his person. "Shall one man's threats," he asked with indignation, "drive me to annul the decrees and institutes of the holy fathers?". . .

On the first of August [1107] a council of the bishops, abbots and chief men of the realm was held at London in the King's palace, and for three whole days the question of the investitures of churches was discussed by the King and the bishops in Anselm's absence, some urging upon him to maintain the practice of his father and brother in defiance of the apostolic command. For the Pope, taking a firm stand upon the decree which had been published thereon, had conceded the homage which Pope Urban had prohibited equally with investitures, and thereby secured the King's consent to his view of the investitures. . . .

Afterwards, in the presence of Anselm, and the whole council standing, the King agreed and ordained that from that time forward no man should be invested in any bishopric or abbey by the King or the hand of any layman in England by the giving of the pastoral staff or ring, Anselm on his side granting that no man elected to be a prelate should be deprived of consecration to the dignity he had received, by reason of the homage which he should do to the King. Upon this settlement of the dispute, institutions were made by the King, without investiture of the pastoral staff or ring, by the counsel of Anselm and the chief men of the realm, to almost all the churches of England, so long bereft of their pastors. . . .

44. The Constitutions of Clarendon, 1164

To regulate relations between civil and ecclesiastical authorities, allegedly in accordance with long established custom, provisions known as the Constitutions of Clarendon were drawn up by a Great Council in 1164. Of particular importance is the third section, relating to clerics against whom charges were made (criminous clerks), and those requiring royal permission for the excommunication of tenant-in-chief or royal servant, for clergy to leave the realm, and for appeals to be carried to Rome. Thomas Becket, Archbishop of Canterbury, opposed the Constitutions, and a bitter quarrel between him and Henry ensued.

In the year 1164 . . . , being . . . the tenth of Henry II, most illustrious King of the English, in the presence of the said King was made this record and declaration of a certain part of the customs, liberties and privileges

---

SOURCE: William Stubbs, *Select Charters and Other Illustrations of English Constitutional History* (Oxford: The Clarendon Press, 1895), 137–40, in *English Historical Documents, 1042–1189,* David C. Douglas and George W. Greenaway (eds.), David C. Douglas (gen. ed.), (London: Eyre and Spottiswoode, 1953; New York: Oxford University Press), 718–22.

of his ancestors, that is, of King Henry his grandfather, and of other things which ought to be observed and maintained in the realm. And by reason of the dissension and discords which had arisen between the clergy and the justices of the lord King and the barons of the realm concerning the customs and privileges of the realm, this declaration was made in the presence of the archbishops, bishops and clergy, and of the earls, barons and magnates of the realm. And these same customs were acknowledged by the archbishops and bishops, and the earls, barons, nobles and elders of the realm. . . .

1. If a dispute shall arise between laymen, or between clerks[13] and laymen, or between clerks, concerning advowson and presentation to churches, let it be treated and concluded in the court of the lord King.

3. Clerks cited and accused of any matter shall, when summoned by the King's justice, come before the King's court to answer there concerning matters which shall seem to the King's court to be answerable there, and before the ecclesiastical court for what shall seem to be answerable there, but in such a way that the justice of the King shall send to the court of holy Church to see how the case is there tried. And if the clerk be convicted or shall confess the Church ought no longer to protect him.

4. It is not lawful for archbishops, bishops and beneficed clergy of the realm to depart from the kingdom without the lord King's leave. And if they do so depart, they shall, if the King so please, give security that neither in going nor in tarrying nor in returning will they contrive evil or injury against the King or the kingdom.

7. No one who holds of the King in chief nor any of the officers of his demesne shall be excommunicated, nor the lands of any one of them placed under interdict, unless applications shall first be made to the lord King, if he be in the realm, or to his chief justice, if he be abroad, that right may be done him; in such wise that matters pertaining to the royal court shall be concluded there and matters pertaining to the ecclesiastical court shall be sent thither to be dealt with.

8. With regard to appeals, if they should arise, they should proceed from the archdeacon to the bishop and from the bishop to the archbishop. And if the archbishop should fail to do justice, the case must finally be brought to the lord King, in order that by his command the dispute may be determined in the archbishop's court, in such wise that it may proceed no further without the assent of the lord King.

9. If a dispute shall arise between a clerk and a layman . . . in respect of any holding which the clerk desires to treat as free alms,[14] but the layman as lay fee, it shall be determined by the recognition of twelve

[13] Clerics.

[14] Frankalmoin tenure, by which land was held in return for the performance of spiritual services.

lawful men through the deliberation, and in the presence of the King's chief justice, whether the holding pertains to free alms or to lay fee.[15] And if it be judged to pertain to free alms, the plea shall be heard in the ecclesiastical court; but if to lay fee, it shall be heard in the King's court, unless both of them shall claim from the same bishop or baron. . . .

11. Archbishops, bishops and all beneficed clergy of the realm, who hold of the King in chief, have their possessions from the lord King by barony and are answerable for them to the King's justices and officers; they observe and perform all royal rights and customs and, like other barons, ought to be present at the judgments of the King's court together with the barons, until a case shall arise of judgment concerning mutilation or death.

12. When an archbishopric or bishopric is vacant, or any abbey or priory of the King's demesne, it ought to be in his own hand, and he shall receive from it all revenue and profits as part of his demesne. And when the time has come to provide for the church, the lord King ought to summon the more important of the beneficed clergy of the church, and the election ought to take place in the lord King's chapel with the assent of the lord King and the advice of the clergy of the realm whom he shall summon for this purpose. And the clerk elected there shall do homage and fealty to the lord King as his liege lord for his life and limbs and his earthly honor, saving his order,[16] before he is consecrated.

13. If any of the magnates of the realm should forcibly prevent an archbishop or bishop or archdeacon from doing justice to himself or to his people, the lord King ought to bring him to justice. And if perchance anyone should forcibly dispossess the lord King of his right, the archbishops, bishops and archdeacons ought to bring him to justice, so that he may make satisfaction to the lord King.

16. Sons of villeins ought not to be ordained without the consent of the lord on whose land they are known to have been born. . . .

45. The Murder of Thomas Becket, 1170

Probably no crime committed in medieval England was regarded as more heinous than the murder of Thomas Becket, Archbishop of

---

SOURCE: J. C. Robertson (ed.), "Edward Grim's Life of Becket," *Materials for the History of Thomas Becket* (London: Rolls Series, 1875–85), *II*, 435–38, in Edward P. Cheyney, *Readings in English History Drawn from the Original Sources* (Boston: Ginn, 1935), 155–58.

---

[15] I.e., whether the land is held by a spiritual or secular tenure.
[16] Saving the rights of his religious order.

Canterbury, carried out by henchmen of Henry II. The results were consequential. In a sense, Henry was responsible for it, since he had by rash words led the actual murderers to believe he desired it. Thomas was regarded as a martyr and within a few years was canonized; Henry did formal penance, and, with public opinion as it was, found it impossible to implement all provisions of the Constitutions of Clarendon, and particularly that restricting the right of benefit of clergy. We are fortunate in having an eye-witness account of the murder, from Edward Grim, an attendant of Becket, in whose arms he died.

When the monks entered the church the four knights followed immediately behind with rapid strides. With them was a certain sub-deacon, armed with malice like their own, Hugh, fitly surnamed for his wickedness, Mauclerc, who showed no reverence for God or the saints, as the result showed. When the holy archbishop entered the church the monks stopped vespers which they had begun and ran to him, glorifying God that they saw their father, whom they had heard was dead, alive and safe. They hastened, by bolting the doors of the church, to protect their shepherd from the slaughter. But the champion, turning to them, ordered the church doors to be thrown open, saying: "It is not meet to make a fortress of the house of prayer, the church of Christ: though it be not shut up it is able to protect its own; and we shall triumph over the enemy rather in suffering them in fighting, for we came to suffer, not to resist." And straightway they entered the house of peace and reconciliation with swords sacrilegiously drawn, causing horror to the beholders by their very looks and the clanging of their arms.

All who were present were in tumult and fright, for those who had been singing vespers now ran hither to the dreadful spectacle.

Inspired by fury the knights called out, "Where is Thomas Becket, traitor to the King and realm?" As he answered not, they cried out the more furiously, "Where is the Archbishop?" At this, intrepid and fearless (as it is written, "The just, like a bold lion, shall be without fear"), he descended from the stair where he had been dragged by the monks in fear of the knights, and in a clear voice answered: "I am here, no traitor to the King, but a priest. Why do ye seek me?" And whereas he had already said that he feared them not, he added, "So I am ready to suffer in His name, who redeemed me by His blood; be it far from me to flee from your swords or to depart from justice." Having thus said, he turned to the right, under a pillar, having on one side the altar of the Blessed Mother of God and ever Virgin Mary, on the other that of St. Benedict the Confessor, by whose example and prayers, having crucified the world with its lusts, he bore all that the murderers could do, with such constancy of soul as if he had been no longer in the flesh.

The murderers followed him. "Absolve," they cried, "and restore to communion those whom you have excommunicated, and their powers to those whom you have suspended." He answered, "There has been no satisfaction, and I will not absolve them." "Then you shall die," they cried, "and receive what you deserve." "I am ready," he replied, "to die for my Lord, that in my blood the church may obtain liberty and peace. But in the name of Almighty God I forbid you to hurt my people, whether clerk or lay." Thus piously and thoughtfully did the noble martyr provide that no one near him should be hurt or the innocent be brought to death, whereby his glory should be dimmed as he hastened to Christ. Thus did it become the martyr knight to follow in the footsteps of this Captain and Saviour, who, when the wicked sought Him, said, "If ye seek me, let these go their way."

Then they laid sacrilegious hands on him, pulling and dragging him that they might kill him outside the church, or carry him away a prisoner, as they afterwards confessed. But when he would not be forced away from the pillar, one of them pressed on him and clung to him more closely. Him he pushed off, calling him "pander," and saying, "Touch me not, Reginald; you owe me fealty and subjection; you and your accomplices act like madmen." The knight, fired with terrible rage at this severe rebuke, waved his sword over the sacred head. "No faith," he cried, "nor subjection do I owe you against my fealty to my lord the King." Then the unconquered martyr, seeing the hour at hand which should put an end to this miserable life, and give him straightway the crown of immortality promised by the Lord, inclined his head as one who prays, and, joining his hands, lifted them up and commended his cause and that of the church to God, to St. Mary, and to the blessed martyr Denys.

Scarce had he said the words when the wicked knight, fearing lest the Archbishop should be rescued by the people and escape alive, leapt upon him suddenly and wounded this lamb who was sacrificed to God, on the head, cutting off the top of the crown which the sacred unction of the chrism had dedicated to God; and by the same blow he wounded the arm of him who tells this. For he, when the others, both monks and clerks, fled, stuck close to the sainted Archbishop and held him in his arms till the arm he interposed was almost severed.

Behold the simplicity of the dove, the wisdom of the serpent, in the martyr who opposed his body to those who struck, that he might preserve his head, that is, his soul and the church, unharmed; nor would he use any forethought against those who destroyed the body whereby he might escape. O worthy shepherd, who gave himself so boldly to the wolves that his flock might not be torn! Because he had rejected the world, the world in wishing to crush him unknowingly exalted him. Then he received a second blow on the head, but still stood firm. At the third blow he fell on his knees and elbows, offering himself a living victim, and saying in a low voice, "For the name of Jesus and the protection of the church I am ready to embrace death." Then the third knight inflicted a terrible wound as he lay, by which the sword was broken against the pavement, and the crown, which was large, was separated from the head; so that the blood white with the brain, and the brain red with blood, dyed the surface of the virgin mother church with the life and death of the confessor and martyr in the colors of the lily and the rose.

The fourth knight prevented any from interfering, so that the others might freely perpetrate the murder. In order that a fifth blow might not be wanting to the martyr who was in other things like to Christ, the fifth (no knight, but that clerk who had entered with the knights) put his foot on the neck of the holy priest and precious martyr, and, horrible to say, scattered his brains and blood over the pavement, calling out to the others, "Let us away, knights; he will rise no more."

SECTION F

The Opposition to King John

46. John's Dispute with the Church

A bitter conflict between John and the Pope, lasting from 1205 to 1213, grew out of a disputed election to the Archbishopric of Canterbury. We are reminded of the lay investiture controversy of a century earlier. The circumstances are thus described by the chronicler, Roger of Wendover (d. 1236), a monk of St. Albans.

About this time the monks of the church of Canterbury appeared before our lord the Pope, to plead a disgraceful dispute which had arisen between themselves; for a certain part of them, by authenticated letters of the convent, presented Reginald, Sub-prior of Canterbury, as they had often done, to be archbishop-elect, and earnestly required the confirmation of his election; the other portion of the same monks had, by letters alike authentic, presented John, Bishop of Norwich, showing by many arguments that the election of the Sub-prior was null, not only because

SOURCE: Roger of Wendover, *Flowers of History,* J. A. Giles (ed.), (London: G. Bell and Sons, 1849), *II,* 236–40.

it had been made by night, and without the usual ceremonies, and without the consent of the King. . . . At length, after long arguments on both sides, our lord the Pope, seeing that the parties could not agree in fixing on the same person, and that both elections had been made irregularly, and not according to the decrees of the holy canons, by the advice of his cardinals annulled both elections, laying the apostolic interdict on the parties, and by definitive judgment ordering that neither of them should again aspire to the honors of the archbishopric.

The aforesaid elections being thus annulled, our lord the Pope, being unwilling to permit the Lord's flock to be any longer without the care of a pastor, persuaded the monks of Canterbury, who had appeared before him as pleaders in the matter of the church of Canterbury, to elect master Stephen Langton, a cardinal priest, a man, as we have said, skilled in literary science, and discreet and accomplished in his manners; and he asserted that the promotion of that person would be of very great advantage, as well to the King himself, as to the whole English church. . . .

When at length the letters of our lord the Pope came to the notice of the English King, he was exceedingly enraged, as much at the promotion of Stephen Langton, as at the annulling of the election of the Bishop of Norwich, and accused the monks of Canterbury of treachery; for he said that they had, to the prejudice of his rights, elected their Sub-prior without his permission, and afterwards, to palliate their fault by giving satisfaction to him, they chose the Bishop of Norwich; that they had also received money from the treasury for their expenses in obtaining the confirmation of the said Bishop's election from the apostolic see; and to complete their iniquity, they had there elected Stephen Langton, his open enemy, and had obtained his consecration to the archbishopric.

On this account the said King, in the fury of his anger and indignation, sent Fulk de Cantelu and Henry de Cornhill, two most cruel and inhuman knights, with armed attendants, to expel the monks of Canterbury, as if they were guilty of a crime against his injured majesty from England, or else to consign them to capital punishment. These knights were not slow to obey the commands of their lord, but set out for Canterbury, and, entering the monastery with drawn swords, in the King's name fiercely ordered the Prior and monks to depart immediately from the kingdom of England as traitors to the King's Majesty; and they affirmed with an oath that, if they (the monks) refused to do this, they would themselves set fire to the monastery, and the other offices adjoining it, and would burn all the monks themselves with their buildings. The monks, acting unadvisedly, departed without violence or laying hands on anyone; all of them, except thirteen sick men who were lying in the infirmary unable to walk, forthwith crossed into Flanders, and were honorably received at the Abbey of St. Bertinus and other monasteries on the continent. . . .

47. John's Homage to the Pope

Fearing an invasion of England by Philip Augustus of France, with whom John (like Henry II and Richard I) was at war, and that he could not rely on his barons to defend the realm in such a case, John in 1213 came to terms with Pope Innocent, doing homage to the latter for both England and Ireland.

John, by the grace of God King of England, Lord of Ireland, Duke of Normandy and Aquitaine, Count of Anjou, to all the faithful in Christ who shall inspect this present charter, greeting.

We will it to be known by all of you by this our charter, confirmed by our seal, that we, having offended God and our mother the holy Church in many things, and being on that account known to need the divine mercy, and unable to make any worthy offering for the performance of due satisfaction to God and the Church, unless we humble ourselves and our realms – we, willing to humble ourselves for Him who humbled Himself for us even to death, by the inspiration of the Holy Spirit's grace, under no compulsion of force or of fear, but of our good and free will, and by the common consent of our barons, offer and freely grant to God and His holy apostles Peter and Paul, and the holy Roman Church, our mother, and to our lord the Pope Innocent and his catholic successors, the whole realm of England and the whole realm of Ireland with all their rights and appurtenances, for the remission of our sins and those of all our race, as well quick as dead; and from now receiving back and holding these, as a feudal dependant, from God and the Roman Church, in the presence of the prudent man Pandulf, subdeacon and familiar of the lord the Pope, do and swear fealty for them to the aforesaid our lord the Pope Innocent and his catholic successors and the Roman Church, according to the form written below, and will do liege homage to the same lord the Pope in his presence if we shall be able to be present before him; binding our successors and heirs by our wife, forever, that in like manner to the supreme pontiff for the time being, and to the Roman Church, they should pay fealty and acknowledge homage without contradiction. Moreover, in proof of this our perpetual obligation and grant, we will and establish that from the proper and special revenues of our realms aforesaid, for all service and custom that we should render for ourselves, saving in all respects the penny of blessed Peter, the Roman Church receive 1000 marks sterling each year, to wit at the feast of St. Michael 500 marks, and at Easter 500 marks;

SOURCE: Thomas Rymer, *Foedera* (London: Record Commission, 1816), *I.* pt. i, 111–12, in R. Trevor Davies, *Documents Illustrating the History of Civilization in Medieval England, 1066–1500* (London: Methuen, 1926), 94–95.

700 to wit for the realm of England, and 300 for the realm of Ireland; saving to us and our heirs our rights, liberties and royalties. All which, as aforesaid, we willing them to be perpetually ratified and confirmed, bind ourselves and our successors not to contravene. And if we or any of our successors shall presume to attempt this, whoever he be, unless he come to amendment after due admonition, let him forfeit right to the kingdom, and let this charter of obligation and grant on our part remain in force forever.

The Oath of Fealty – I, John, by the grace of God King of England and Lord of Ireland, from this hour forward will be faithful to God and the blessed Peter and the Roman Church, and my lord the Pope Innocent and his successors following in catholic manner: I will not be party in deed, word, consent, or counsel, to their losing life or limb or being unjustly imprisoned. Their damage, if I am aware of it, I will prevent, and will have removed if I can; or else, as soon as I can, I will signify it, or will tell such persons as I shall believe will tell them certainly. Any counsel they entrust to me, immediately or by their messengers or their letter, I will keep secret, and will consciously disclose to no one to their damage. The patrimony of blessed Peter, and specially the realm of England and the realm of Ireland, I will aid to hold and defend against all men to my ability. So help me God and these holy gospels. Witness myself at the house of the Knights of the Temple near Dover, in the presence of the lord H. Archbishop of Dublin; the lord J. Bishop of Norwich; G. Fitz-Peter, Earl of Essex, our Justiciar; W. Earl of Salisbury, our brother; W. Marshall, Earl of Pembroke; R. Count of Boulogne; E. Earl of Warenne; S. Earl of Winchester; W. Earl of Arundel; W. Earl of Ferrers; W. Brewer; Peter, son of Herbert; Warren, son of Gerald. The 15th day of May in the 14th year of our reign.

48. John's Charter to the Church, 1214

In 1214, with opposition at the point of rebellion, King John sought to detach the clergy from the insurgent barons by granting to the church a charter guaranteeing nonintervention in the election of its bishops:

John, by the grace of God King of England, Lord of Ireland, Duke of Normandy and of Aquitaine, and Count of Anjou, to archbishops, bishops,

---

SOURCE: William Stubbs, *Select Charters and Other Illustrations of English Constitutional History* (Oxford: The Clarendon Press, 1895), 288–89, in Carl Stephenson and Frederick G. Marcham, *Sources of English Constitutional History* (New York: Harper and Row, 1937), 114–15.

earls, barons, knights, bailiffs, and all who may hear or see these letters, greeting. Since, by the grace of God, a full agreement with regard to damages and usurpations during the time of the interdict has been established, of the pure and free will of each party, between us and our venerable fathers . . . we wish, not merely to give them satisfaction to the best, God willing, of our ability, but also to make sound and useful provision for the whole English Church in perpetuity. Accordingly, no matter what sort of custom has hitherto been observed in the English Church, either in our time or in that of our predecessors, and no matter what right we have hitherto asserted for ourself in the election of any prelates, we [now], on the petition of those [prelates], for the health of our soul and [the souls] of our predecessors and successors, kings of England, of our own pure and free will and by the common assent of our barons, have granted and established and by this our charter have confirmed that in all and singular of the churches, monasteries, cathedrals and convents of our whole realm of England the elections of whatsoever prelates, both greater and lesser, shall henceforth and forever be free, saving to us and our heirs the custody of vacant churches and monasteries that belong to us. We also promise that we will not hinder, nor will we permit or authorize our men to hinder, the electors in any or all of the aforesaid churches and monasteries, when prelacies become vacant, from freely appointing a pastor over themselves whenever they please, providing, however, that permission to elect has first been sought from us and our heirs – which we will not deny or delay. And if perchance – which God forbid! – we should deny or delay [permission], the electors shall nevertheless proceed to make a canonical election. Moreover, after the election has been held, our confirmation is to be requested; which likewise we will not deny, unless we can bring forward and lawfully prove some reasonable cause for which we ought not to give confirmation. Wherefore we will and straitly enjoin that, when churches or monasteries are vacant, no one shall act or presume to act in any way contrary to this our grant and constitution. If, however, any one at any time or in any way shall act contrary to it, may he incur the malediction of Almighty God and our own! . . .

49. The Winning of Magna Carta

Though John had made his peace with the church, and thus had the Pope on his side, he still had to contend with the rebellious barons, who were made even more critical of the regime after the disastrous

---

SOURCE: Roger of Wendover, *Flowers of History*, J. A. Giles (ed.), (London: G. Bell and Sons, 1849), *II*, 304–09.

campaigns of 1214. The chronicler, Roger of Wendover (d. 1236) gives an account of events in 1215 leading to the formulation of Magna Carta.

A.D. 1215, which was the seventeenth year of the reign of King John; he held his court at Winchester at Christmas for one day, after which he hurried to London, and took up his abode at the New Temple; and at that place the above-mentioned nobles came to him in gay military array, and demanded the confirmation of the liberties and laws of King Edward, with other liberties granted to them and to the kingdom and church of England, as were contained in the charter and . . . laws of Henry the First. They also asserted that, at the time of his absolution at Winchester, he had promised to restore those laws and ancient liberties, and was bound by his own oath to observe them. The King, hearing the bold tone of the barons in making this demand, much feared an attack from them, as he saw that they were prepared for battle; he however made answer that their demands were a matter of importance and difficulty, and he therefore asked a truce till the end of Easter, that he might, after due deliberation, be able to satisfy them as well as the dignity of his crown.

After much discussion on both sides, the King at length, although unwillingly, procured the Archbishop of Canterbury, the Bishop of Ely, and William Marshal, as his sureties, that on the day pre-agreed on he would, in all reason, satisfy them all, on which the nobles returned to their homes. The King however, wishing throughout England to swear fealty to him alone against all men, and to renew their homage to him; and, the better to take care of himself, he, on the day of St. Mary's purification,[17] assumed the cross of our Lord,[18] being induced to this more by fear than devotion. . . .

In Easter week of this same year, the above-mentioned nobles assembled at Stamford, with horses and arms; for they had now induced almost all the nobility of the whole kingdom to join them, and constituted a very large army; for in their army there were computed to be two thousand knights, besides horse soldiers, attendants, and foot soldiers, who were variously equipped. . . . The King at this time was awaiting the arrival of his nobles at Oxford. On the Monday next after the octaves of Easter, the said barons assembled in the town of Brackley: and when the King learned this, he sent the Archbishop of Canterbury, and William Marshal, Earl of Pembroke, with some other prudent men, to them to inquire what the laws and liberties were which they demanded. The barons then delivered to the messengers a paper, containing in great measure the laws

[17] February 2.
[18] As a crusader.

and ancient customs of the kingdom, and declared that, unless the King immediately granted them and confirmed them under his own seal, they would, by taking possession of his fortresses, force him to give them sufficient satisfaction as to their before-named demands. The Archbishop with his fellow messengers then carried the paper to the King, and read to him the heads of the paper one by one throughout. The King when he heard the purport of these heads, derisively said, with the greatest indignation, "Why, amongst these unjust demands, did not the barons ask for my kingdom also? Their demands are vain and visionary, and are unsupported by any plea of reason whatever." And at length he angrily declared with an oath, that he would never grant them such liberties as would render him their slave. . . .

As the Archbishop and William Marshal could not by any persuasions induce the King to agree to their demands, they returned by the King's order to the barons, and duly reported all they had heard from the King to them; and when the nobles heard what John said, they appointed Robert Fitz-Walter commander of their soldiers, giving him the title of "Marshal of the army of God and the holy church," and then, one and all flying to arms, they directed their forces towards Northampton. On their arrival there they at once laid siege to the castle, but after having stayed there for fifteen days, and having gained little or no advantage, they determined to move their camp; for having come without petrariae[19] and other engines of war, they, without accomplishing their purpose, proceeded in confusion to the castle of Bedford. . . .

When the army of the barons arrived at Bedford, they were received with all respect by William de Beauchamp. There also came to them there messengers from the city of London, secretly telling them, if they wished to get into that city, to come there immediately. The barons, inspirited by the arrival of this agreeable message, immediately moved their camp and arrived at Ware; after this they marched the whole night, and arrived early in the morning at the city of London, and, finding the gates open, they, on the 24th of May, which was the Sunday next before our Lord's ascension, entered the city without any tumult whilst the inhabitants were performing divine service; for the rich citizens were favorable to the barons, and the poor ones were afraid to murmur against them. The barons having thus got into the city, placed their own guards in charge of each of the gates, and then arranged all matters in the city at will. They then took security from the citizens, and sent letters throughout England to those earls, barons, and knights, who appeared to be still faithful to the King, though they only pretended to be so, and advised them with threats, as they regarded the safety of all their property and

[19] Petraries, machines for hurling large stones.

possessions, to abandon a king who was perjured and who warred against his barons, and together with them to stand firm and fight against the King for their rights and for peace; and that, if they refused to do this, they, the barons, would make war against them all, as against open enemies, and would destroy their castles, burn their houses and other buildings, and destroy their warrens, parks, and orchards. . . .

King John, when he saw that he was deserted by almost all, so that out of his regal superabundance of followers he scarcely retained seven knights, was much alarmed lest the barons would attack his castles and reduce them without difficulty, as they would find no obstacle to their so doing; and he deceitfully pretended to make peace for a time with the aforesaid barons, and sent William Marshal, Earl of Pembroke, with other trustworthy messengers, to them, and told them that, for the sake of peace, and for the exaltation and honor of the kingdom, he would willingly grant them the laws and liberties they required; he also sent word to the barons by these same messengers, to appoint a fitting day and place to meet and carry all these matters into effect. The King's messengers then came in all haste to London, and without deceit reported to the barons all that had been deceitfully imposed on them; they in their great joy appointed the fifteenth of June for the King to meet them, at a field lying between Staines and Windsor. Accordingly, at the time and place pre-agreed on, the King and nobles came to the appointed conference, and when each party had stationed themselves apart from the other, they began a long discussion about terms of peace and the aforesaid liberties. . . .

50. Magna Carta, 1215

Magna Carta (the Great Charter) represents concessions that King John was forced to make to his barons, and to some extent to his other free subjects. It is a highly practical document, dealing with contemporary affairs, and not a blueprint for democracy or a philosophical expression of human rights. Yet it continues to be regarded as one of England's constitutional landmarks, in that the concessions made by John, and confirmed by later monarchs, proclaim the superiority of law over the will of the king. The Charter was revised in 1216 and 1217, and took final form in 1225. In the following extracts, passages omitted in these revisions are printed in italics.

SOURCE: William Stubbs, *Select Charters and Other Illustrations of English Constitutional History* (Oxford: The Clarendon Press, 1895), 296–306, in Carl Stephenson and Frederick G. Marcham, *Sources of English Constitutional History* (New York: Harper and Row, 1937), 115–26.

John, by the grace of God King of England, . . . to his archbishops, bishops, abbots, earls, barons, justiciars,[20] *foresters, sheriffs, reeves, ministers, and all his bailiffs and faithful men, greeting. Know that, through the inspiration of God, for the health of our soul and [the souls] of all our ancestors and heirs, for the honor of God and the exaltation of Holy Church, and for the betterment of our realm, by the counsel of our venerable fathers . . . , of our nobles . . . , and of our other faithful men:*

1. We have in the first place granted to God and by this our present charter have confirmed, for us and our heirs forever, that the English church shall be free and shall have its rights entire and its liberties inviolate. . . . We have also granted to all freemen of our kingdom, for us and our heirs forever, all the liberties hereinunder written, to be had and held by them and their heirs of us and our heirs.

2. If any one of our earls or barons or other men holding of us in chief dies, and if when he dies his heir is of full age and owes relief, [that heir] shall have his inheritance for the ancient relief . . .

3. If, however, the heir of any such person is under age and is in wardship, he shall, when he comes of age, have his inheritance without relief and without fine.

4. The guardian of the land of such an heir who is under age shall not take from the land of the heir more than reasonable issues and reasonable customs and reasonable services, and this without destruction and waste of men or things. . . .

6. Heirs shall be married without disparagement; *yet so that, before the marriage is contracted, it shall be announced to the blood relatives of the said heir.*

7. A widow shall have her marriage portion and inheritance immediately after the death of her husband . . . ; nor shall she give anything for her dowry or for her marriage portion or for her inheritance . . .

8. No widow shall be forced to marry so long as she wishes to live without a husband; yet so that she shall give security against marrying without our consent if she holds of us, or without the consent of her lord if she holds of another.

9. Neither we nor our bailiffs will seize any land or revenue for any debt, so long as the chattels of the debtor are sufficient to repay the debt . . .

12. *Scutage or aid shall be levied in our kingdom only by the common counsel of our kingdom, except for ransoming our body, for knighting our eldest son, and for once marrying our eldest daughter; and for these*

[20] Judges.

[purposes] only a reasonable aid shall be taken. The same provision shall hold with regard to the aids of the city of London.

13. And the city of London shall have all its ancient liberties and free customs, *both by land and by water.* Besides, we will and grant that all the other cities, boroughs, towns and ports shall have all their liberties and free customs.

14. *And in order to have the common counsel of the kingdom for assessing aid other than in the three cases aforesaid, or for assessing scutage, we will cause the archbishops, bishops, abbots, earls and greater barons to be summoned by our letters individually; and besides we will cause to be summoned in general, through our sheriffs and bailiffs, all those who hold of us in chief—for a certain day, namely, at the end of forty days at least, and to a certain place. And in all such letters of summons we will state the cause of the summons; and when the summons has thus been made, the business assigned for the day shall proceed according to the counsel of those who are present, although all those summoned may not come.*

15. *In the future we will not grant to anyone that he may take aid from his freemen, except for ransoming his body, for knighting his eldest son, and for once marrying his eldest daughter; and for these [purposes] only a reasonable aid shall be taken.*

16. No one shall be distrained to render greater service from a knight's fee, of from any other free tenement, than is thence owed.

17. Common pleas shall not follow our court, but shall be held in some definite place.

20. A freeman shall be amerced for a small offence only according to the degree of the offence; and for a grave offence he shall be amerced according to the gravity of the offense, saving his contenement.[21] And a merchant shall be amerced in the same way, saving his merchandise; and a villein in the same way, saving his wainage[22]—should they fall into our mercy. And none of the aforesaid amercements shall be imposed except by the oaths of good men from the neighborhood.

21. Earls and barons shall be amerced only by their peers, and only according to the degree of the misdeed.

24. No sheriff, constable, coroner or other bailiff of ours shall hold the pleas of our crown.

28. No constable or other bailiff of ours shall take grain or other chattels of anyone without immediate payment therefor in money, unless by the will of the seller he may secure postponement of that [payment].

29. No constable shall distrain any knight to pay money for castle-

[21] That which is sufficient to maintain a livelihood.
[22] Horses, oxen and other means of tillage.

guard when he is willing to perform that service himself, or through another good man if for reasonable cause he is unable to perform it himself. And if we lead or send him on a military expedition, he shall be quit of [castle-]guard for so long a time as he shall be with the army *at our command.*

32. We will hold the lands of those convicted of felony only for a year and a day, and the lands shall then be given to the lords of the fiefs [concerned].

34. Henceforth the writ called praecipe[23] shall not be issued for anyone concerning any tenement whereby a freeman may lose his court.

35. There shall be one measure of wine throughout our entire kingdom, and one measure of ale; also one measure of grain, namely, the quarter of London; and one width of dyed cloth, russet [cloth], and hauberk [cloth], namely, two yards between the borders. With weights, moreover, it shall be as with measures.

39. No freeman shall be captured or imprisoned or disseised[24] or outlawed or exiled or in any way destroyed, nor will we go against him or send against him, except by the lawful judgment of his peers or by the law of the land.

40. To no one will we sell, to no one will we deny or delay right or justice.

41. All merchants may safely and securely go away from England, come to England, stay in and go through England, by land or by water, for buying and selling under right and ancient customs and without any evil exactions, except in time of war if they are from the land at war with us . . .

45. *We will appoint as justiciars, constables, sheriffs, or bailiffs only such men as know the law of the kingdom and well desire to observe it.*

52. *If anyone, without the lawful judgment of his peers, has been disseised or deprived by us of his lands, castles, liberties, or rights, we will at once restore them to him. And if a dispute arises in this connection, then let the matter be decided by the judgment of the twenty-five barons, concerning whom provision is made below . . .*

55. *All fines which have been made with us unjustly and contrary to the law of the land, and all amercements* [*so*] *made . . . are to be entirely pardoned; or decision is thereon to be made by the judgment of the twenty-five barons concerning whom provision is made below . . . , or by the judgment of the majority of them, together with . . . Stephen, Archbishop of Canterbury, if he can be present, and other men whom he may wish to associate with himself for this purpose . . .*

23 A writ commanding sheriffs to send certain cases for trial in the king's courts.
24 Dispossessed.

60. Now all these aforesaid customs and liberties, which we have granted, in so far as concerns us, to be observed in our kingdom toward our men, all men of our kingdom, both clergy and laity, shall, in so far as concerns them, observe toward their men.

61. *Since moreover for [the love of] God, for the improvement of our kingdom, and for the better allayment of the conflict that has arisen between us and our barons, we have granted all these [liberties] aforesaid, wishing them to enjoy those [liberties] by full and firm establishment forever, we have made and granted them the following security: namely, that the barons shall elect twenty-five barons of the kingdom, whomsoever they please, who to the best of their ability should observe, hold, and cause to be observed the peace and liberties that we have granted to them and have confirmed by this our present charter; so that, specifically, if we or our justiciar or our bailiffs or any of our ministers are in any respect delinquent toward anyone or transgress any article of the peace or the security, and if the delinquency is shown to four barons of the aforesaid twenty-five barons, those four barons shall . . . explain to us the wrong, asking that without delay we cause this wrong to be redressed. And if within a period of forty days . . . we do not redress the wrong, . . . the four barons aforesaid shall refer that case to the rest of the twenty-five barons, and those twenty-five barons, together with the community of the entire country, shall distress and injure us in all ways possible – namely, by capturing our castles, lands and possessions and in all ways that they can – until they secure redress according to their own decision, saving our person and [the person] of our queen and [the persons] of our children. And when redress has been made, they shall be obedient to us as they were before. And any one in the land who wishes shall swear that, for carrying out the aforesaid matters, he will obey the commands of the twenty-five barons aforesaid and that he, with his men, will injure us to the best of his ability; and we publicly and freely give license of [thus] swearing to everyone who wishes to do so . . . Moreover, all those of the land who of themselves and by their own free will are unwilling to take the oath for the twenty-five barons, with them to distress and injure us, we will by our mandate cause to swear [such an oath] as aforesaid . . .*

By the witness of the aforesaid men and of many others. Given by our hand in the meadow that is called Runnymede between Windsor and Staines, June 15, in the seventeenth year of our reign.

SECTION G

Complaint and Reform under Henry III

51. Opposition to Henry III

The dissatisfaction of the barons did not cease with the granting of Magna Carta. John's reign ended in war between king and barons, and the long reign of John's son, Henry III (1216–1272) provided many occasions for baronial criticism. The chronicler Matthew Paris (d. 1259) tells us of complaints made to the King at a great council held early in 1248.

Early in the course of that year, namely on the octave of the Purification,[25] the nobility of the whole realm of England was summoned together by royal edict, and met in London to treat with the lord King diligently and effectually concerning the affairs of the realm, its excessive disturbance and poverty, and its shameful loss of strength in our time. Therefore there gathered together there not only a great crowd of barons, knights and nobles, with abbots, priors and clergy, but also nine bishops with as many earls, namely the Archbishop of York, the Bishops of Winchester, Lincoln, Norwich, Worcester, Chichester, Ely, Rochester, Carlisle; the Earls of Gloucester, Leicester, Winchester, Hereford, Oxford, Earl Roger Bigod the Marshal, the Earl of Lincoln, the Earl of Ferrers, the Earl of Warenne, the Earl of Richmond, Peter of Savoy. However, there were not present at this great assembly the Archbishop of Canterbury, Boniface, who was fighting for the lord Pope beyond seas, the Bishop of Durham, who was afar off and an invalid, and the Bishop of Bath, for he had died a little before.

And since the lord King had proposed to demand a pecuniary aid (for he did not conceal his purpose from the assembly), he was severely taken to task because he did not blush to ask for such help at such a time; especially because at the last similar exaction to which the English nobles had unwillingly consented, he promised by his charter that he would not afflict his magnates with any more such injuries and oppressions. He was

---

SOURCE: Matthew Paris, *Chronica Majora*, Henry P. Luard (ed.), (London: Rolls Series, 1872–83), V, 5–7, in Margaret A. Hennings, *England under Henry III* (London: Longmans, Green, 1924), 65–67.

---

[25] February 9.

also reprehended very sternly (nor is that wonderful) because of his indiscreet invitations to aliens, to whom he foolishly, incredibly and extravagantly distributed and scattered all the good things of the kingdom. Also he married the nobles of the realm to ignoble foreigners, scorning and setting aside his own native-born and natural subjects, and doing without that finishing touch of marriage, mutual assent. Moreover, he was blamed, and not undeservedly, because whatever he paid out in food, drink and even clothes, and especially in wine, he violently seized, against the will of their rightful owners who were about to sell them. Wherefore the native-born merchants hide themselves, and so do aliens who intended to bring goods to that part of the world for sale, and so ceases commerce by which divers races are mutually supported and enriched. And we are deprived of reputation and impoverished because they bring back nothing from the King but lawsuits and quibbles, wherefore the lord King incurs tremendous revilings from innumerable people to the peril and infamy of himself and the whole realm. Moreover, from these merchants he violently, and without payment or recompense, snatches wax, silken cloths and other things wherewith to make indiscreet alms and immoderate offerings of candles, to the scandal of himself and the kingdom and of all its inhabitants, not without grave offence to God. "I hate robbery for a burnt offering."[26] In all these things he so acts the tyrant and seeks to rage violently that he does not even permit fish to be caught by poor fishers on the seashore where they ply their trade, and they dare not appear on the borders of the sea nor in cities to be robbed, but they judge it safer to commit themselves to the stormy waves and seek further shores. Also the wretched bankers are browbeaten and savagely tormented by royal tax-collectors, and punishment is added to loss and injury is heaped on injury. And they are compelled to carry on their own backs or on wearied horses, to remote places, in stormy weather and over rough roads the very things which are being taken from them. Again, the lord King was reprimanded because bishoprics and abbacies which were founded by our holy and high-souled fathers, and of which he is supposed to be the protector and defender, he holds like vacant wardships for a long time in his grasp, as though they were thus "in his hand," that is, under his protection. Wherein he goes contrary to the oath which he made first and before all at his coronation, for he impoverishes them to their utter ruin. Then the lord King was severely blamed by all and singular and with no small upbraidings, on the ground that, unlike the mighty kings, his predecessors, he has neither justiciar, chancellor nor treasurer appointed by the common advice of the realm as would seem becoming and useful, but such men as follow his wishes whatever they may be,

[26] Isaiah 61: 8.

provided that it is profitable to him, those who seek not the welfare of the realm but only that of individuals, since they first collect money and procure wardships and incomes for themselves.

52. Baronial Reform of the Government, 1258

At a Great Council held in April, 1258, Henry III agreed to "amend the state of the realm." As may be seen below, he promised to cooperate in arrangements for reform to be made by twenty-four magnates, or greater barons, half of them drawn from his Council and the other half made up of other notables. Thus the barons attempted to exercise a more direct control over the royal government, prior to the emergence of a regular parliamentary organization. In the same year, by the ordinance known as the Provisions of Oxford, the government was practically entrusted to fifteen barons. The experiment failed, however, as did a subsequent arrangement for a baronial council of nine, dominated by Simon de Montfort, the leader of the insurgents.

The King to all, etc. Know that we have conceded to the nobles and magnates of our realm, after oath made of our soul through Robert Walerand,[27] that by twelve faithful men of our Council already chosen and by twelve others of our faithful men chosen on behalf of the nobles, who shall meet at Oxford a month from next Whitsuntide,[28] settlement, rectification and reform of the state of our realm shall be undertaken as may seem expedient, for the honor of God, fealty to us and the profit of our kingdom.

And if by chance any of those chosen on our side shall be absent, let it be lawful for those who are present to find substitutes for those absent; and likewise on the side of our said nobles and faithful men.

And whatever shall be settled in this matter through the twenty-four chosen on both sides and sworn for this purpose, or the majority of them, we will inviolably observe, and we wish and henceforward firmly command that their settlement be inviolably observed by all.

And every kind of security which they or the majority of them shall provide for the observance of this matter, we will make and take care to have made for them fully and without contradiction.

SOURCE: Thomas Rymer, *Foedera* (London: Record Commission, 1816), *I*, pt. i, 371, in Margaret A. Hennings, *England under Henry III* (London: Longmans, Green, 1924), 160–61.

---

[27] Justiciar of the realm.
[28] The week after Whitsunday, the seventh Sunday after Easter.

Also we bear witness that Edward our first-born son, by a corporal oath and by his letters, has conceded that all things expressed and granted above he will as far as in him lies observe faithfully and inviolably and will cause to be observed forever.

Also the said earls and barons have promised that when the aforesaid matters have been dealt with, they will labor in good faith to the end that a common aid be granted us by the community of our realm. . . .

SECTION H

Countryman and Townsman

53. Obligations of the Serf

The overwhelming majority of medieval Englishmen were unfree peasants, holding small acreages in return for labor services and payments to their lords. The following statement, applying to a Sussex manor at the beginning of the fourteenth century, illustrates these obligations.

. . . John of Cayworth holds a house and thirty acres of land, and owes yearly 2s at Easter and Michaelmas; and he owes a cock and two hens at Christmas, of the value of 4d.

And he ought to harrow for two days at the Lenten sowing with one man and his own horse and his own harrow, the value of the work being 4d; and he is to receive from the lord on each day three meals, of the value of 5d, and then the lord will be at a loss of 1d. Thus his harrowing is of no value to the service of the lord.

And he ought to carry the manure of the lord for two days with one cart, with his own two oxen, the value of the work being 8d; and he is to receive from the lord each day three meals of the price as above. And thus the service is worth 3d clear.

And he shall find one man for two days for mowing the meadow of the lord, who can mow, by estimation, one acre and a half, the value of the

---

SOURCE: S. R. Scargill-Bird (ed.), *Custumals of Battle Abbey in the Reigns of Edward I and Edward II* (London: Royal Historical Society, 1887), 19–22, in Edward P. Cheyney, *Readings in English History Drawn from the Original Sources* (Boston: Ginn, 1935), 215–17.

mowing of an acre being 6d; the sum is therefore 9d, and he is to receive each day three meals of the value given above; and thus that mowing is worth 4d clear.

And he ought to gather and carry that same hay which he has cut, the price of the work being 3d.

And he shall have from the lord two meals for one man, of the value of 1½d. Thus the work will be worth 1½d clear.

And he ought to carry the hay of the lord for one day with a cart and three animals of his own, the price of the work being 6d. And he shall have from the lord three meals of the value of 2½d. And thus the work is worth 3½d clear.

And he ought to carry in autumn beans or oats for two days with a cart and three animals of his own, the value of a work being 12d. And he shall receive from the lord each day three meals of the value given above; and thus the work is worth 7d clear.

And he ought to carry wood from the woods of the lord as far as the manor house for two days in summer with a cart and three animals of his own, the value of the work being 9d. And he shall receive from the lord each day three meals of the price given above, and thus the work is worth 4d clear.

And he ought to find one man for two days to cut heath,[29] the value of the work being 4d, and he shall have three meals each day of the value given above; and thus the lord will lose, if he receives the service, 3d. Thus that mowing is worth nothing to the service of the lord.

And he ought to carry the heath which he has cut, the value of the day's work being 5d. And he shall receive from the lord three meals at the price of 2½d. And thus the work will be worth 2½d clear.

And he ought to carry to Battle[30] twice in the summer season, each time half a load of grain, the value of the service being 4d. And he shall receive in the manor each time one meal of the value of 2d. And thus the work is worth 2d clear.

The total of the rents, with the value of the hens, is 2s 4d.

The total of the value of the works is 2s 3½d; owed from the said John yearly.

William of Cayworth holds a house and 30 acres of land and owes at Easter and Michaelmas 2s rent. And he shall do all customs just as the aforesaid John of Cayworth. . . .

And it is to be noted that none of the above-named villeins can give their daughters in marriage nor cause their sons to be tonsured,[31] nor can

[29] Heather.
[30] The town of Battle, in Sussex.
[31] I.e., become clerics, the tonsure, or shaven crown, being a distinctive mark of ecclesiastics.

they cut down timber growing on the lands they hold, without license of the bailiff or sergeant of the lord and then for building purposes and not otherwise. And after the death of any one of the aforesaid villeins the lord shall have as a heriot his best animal, if he had any; if, however, he have no living beast, they say that the lord shall have no heriot. The sons or daughters of the aforesaid villeins shall give for entrance into the holding after the death of their predecessors as much as they give of rent per year.

54. The Serf's Legal Status, c. 1190

In his treatise, *De Legibus Angliae,* written about 1190, Glanvill presents a harsh picture of the legal inferiority of the serf. It has been noted, however, that as time went on the rigid conditions referred to below were considerably modified: "the serf's condition seems better described as unprotectedness than as rightlessness."[32]

This must be noted, that no man who is in serfdom can buy his liberty with his own money; for, even if he had paid the price, he might be recalled to villenage by his lord according to the law and custom of this land; for all the chattels of all serfs are understood to be so far within the power of his lord, that he cannot redeem himself from his lord by any money of his own. . . .

And this also, that a man may free his serf, so far as regards his own person or his own heirs, but not as regards others. For if any serf thus freed were brought into court to make suit against any other man, or to make any law of the land, then he might justly be removed, if his former state of serfdom were objected and proved against him; and this would be so, even though the man thus freed from villenage had been dubbed a knight.

Item, if any serf shall have dwelt unclaimed for a whole year and a day in any chartered town, so that he hath been received into the community or gild of that town as a citizen, then that single fact shall free him from villenage.

SOURCE: Glanvill, *De Legibus Angliae,* bk. v, chap. v, in G. G. Coulton, *Social Life in Britain from the Conquest to the Reformation* (London: Cambridge University Press, 1956), 338–39.

---

32 Frederick Pollock and F. W. Maitland, *History of English Law* (Cambridge, 1898), I, 417.

55. The Supervision of Manors

From *Fleta*, a Latin textbook on English law probably written around 1290, come these extracts describing a model manor and how it should be run, the point of view being obviously that of the lord.

Let the lord then procure a Seneschal;[33] a man circumspect and faithful, provident, discreet and gracious, humble and chaste and peaceful and modest, learned in the laws and customs of his province and in the duties of a Seneschal; one who will devote himself to guard his lord's rights in all matters, and who knoweth how to teach and instruct his master's under-bailiffs in their doubts and their errors; merciful to the poor, turning aside from the path of justice neither for prayers nor for bribes. . . . Let the Seneschal see to it that, in every manor, he measures clearly and openly with the common rod how many acres of arable land it may contain, and learn how much seed is needed, of every sort, for the plough-land, lest cunning reeves should reckon too much in their computations. Let him heed again that all offices be securely locked; for an easy access oftentimes tempteth the weaker brethren to sin. . . . Let him also inquire concerning the bailiffs or sergeants of each manor, and of their underlings, how they have borne and demeaned themselves towards their neighbors and the lord's tenants and other folk; and let them be removed from the room during this inquisition, lest the truth be suppressed through fear. Let him inquire whether they have meddled with disseisins[34] of any kind, or blows or scuffles or wrestlings; or if they neglect their duties to haunt taverns and wakes by night, whereby the lord may suffer loss. . . . Again, he must know the wardships and marriages, at what time they fall into his lord's hands, and what is their yearly value. . . . Moreover, he must forbid, both in general and in especial, the flaying of any sheep or other beast until it have been seen by the bailiff and reeve or other trustworthy witnesses competent to judge of the manner of its death, whether it have been slain of set purpose . . . or by chance and evil fortune. . . . Again, he must compute for the household expenses every night, whether in person or by deputy, on the lord's account, with the marketer, the mareschal,[35] the cook, the spencer,[36] and the other officials; and he must

---

SOURCE: John Selden *Ad Fletam Dissertatio* (London: M. F. for W. Lee, 1647), in G. G. Coulton, *Social Life in Britain from the Conquest to the Reformation* (London: Cambridge University Press, 1956), 301–03.

---

[33] Steward.
[34] Unlawful dispossessions.
[35] Chief groom.
[36] One in charge of the provisions in a household.

make out the sum of expenses for that day. So must he also compute with the larderer, according as each kind of flesh or fish be received by tally; and the carving must be done in his presence, and he must reckon up the joints with the cook and receive a reasonable account thereof. He must know exactly how many halfpenny-loaves are to be made from a quarter of corn, that the baker may deliver the due tale to the pantler;[37] also, how many dishes are needed for the household in themselves[38] on common days. . . . And all the servants, severally and collectively, are bound to answer for their offices to the Seneschal, who for his part is bound to report of their behavior to his lord. . . .

The Bailiff of every manor should be truthful in word, diligent and faithful in deed. . . . Let him beware of blame for sloth; therefore let him arise betimes in the morning, lest he seem but lukewarm and remiss. Let him first see to the plough-yoking, and then go round to survey the fields, woods, meadows, and pastures, lest damage be done there at dawn. . . . Let him see that the ploughmen do their work diligently and well; and, so soon as the ploughs be unyoked, let him measure forthwith the work that hath been done that day; for, unless the ploughmen can give reasonable excuse, they are bound to be accountable for the whole day's ploughing. Nevertheless he must watch their labor and their shortcomings over and over again, and make sure through the hayward[39] that such defaults be visited with due correction and punishment.

56. Henry I's Charter to the City of London

In the course of the twelfth and thirteenth centuries, inhabitants of towns made important strides toward self-government. Various rights and privileges were secured from lords through the purchase of charters. Henry I granted the following charter of liberties to the city of London. A. L. Poole notes that this charter was "the first assertion of what came to be the common aspiration of all boroughs – emancipation from the financial and judicial organization of the shire."[40]

Henry, by the grace of God King of the English to the Archbishop of Canterbury and his bishops and abbots and earls and barons and justices

SOURCE: Thomas Rymer, *Foedera* (London: Record Commission, 1816), *I*, pt. i, 11, in A. E. Bland, *The Normans in England, 1066–1154* (London: G. Bell and Sons, 1914), 74–76.

37 One in charge of a pantry.
38 Apart from guests.
39 An official appointed to look after hedges and fences, and to guard the common herd.
40 A. L. Poole, *From Domesday Book to Magna Carta, 1087–1216* (Oxford, 1951), 69.

and sheriffs and all his trusty men, French and English, of the whole of England, greeting. Know ye that I have granted to my citizens of London that they hold Middlesex at farm[41] for £300 at account, to them and their heirs, of me and my heirs, so that the citizens appoint as sheriff whom they choose from among themselves, and as justice whom they choose from among themselves, to keep the pleas of my crown and to hold the same pleas; and no other shall be justice over the same men of London. And the citizens shall not plead outside the walls for any plea, and they shall be quit of scot and of lot,[42] of Danegeld and murder-fine,[43] and none of them shall suffer trial by battle. And if any of the citizens be impleaded of pleas of the crown, a man of London shall make his proof by the oath that shall be adjudged in the city. And within the walls of the city no man shall be lodged either of my household or of another's, unless lodging be delivered to him. And all men of London and all their possessions shall be quit and free, throughout the whole of England and throughout seaports, of toll and passage and lastage[44] and all other customs. And the churches and the barons and the citizens shall have and hold their sokes[45] duly and peaceably with all customs, so that guests lodged in their sokes give their customs to none save to him whose soke it is, or to the minister whom he shall set there. And a man of London shall not be adjudged to a money penalty, except to his "wer,"[46] to wit, 100s; I speak of pleas to which a money penalty is attached. And there shall no longer be miskenning[47] in the husting[48] or in the folkmoot or in other pleas within the city. And the husting shall sit once a week, to wit, on Monday. And I will cause my citizens to have their lands and pledges and debts within the city and without. And I will award them right by the law of the city touching the lands whereto they shall lay claim before me. And if any man take toll or custom from the citizens of London, the citizens of London shall take from the borough or town where the toll or custom was taken as much as the man of London gave by way of toll, and further he shall take his damages. And all debtors who owe debts to the citizens shall render the same to them or shall prove in London that they owe nothing. And if they refuse to render the debts or to bring it to proof, then the citizens to whom their debts are due shall take their pledges within the city or from the county in which the debtor dwells.

[41] The *firma burgi,* or lump-sum payment in return for rights and privileges.
[42] A form of taxation.
[43] The fine imposed in homicides where victim and assailant were unknown.
[44] A duty on freight or transportation.
[45] Jurisdictions.
[46] The blood price, deriving from Anglo-Saxon custom.
[47] A fine for an error in reciting formal oaths of innocence.
[48] A court in London.

And the citizens shall have their hunting chases as well and fully as their ancestors had the same, to wit, Ciltre and Middlesex and Surrey . . .

57. Henry II's Charter to the London Weavers

A prominent feature of medieval economic life was the development of craft guilds. The members of these guilds enjoyed exclusive rights with regard to the manufacture of particular articles. Such rights were frequently established or confirmed by royal decree, as may be seen from the following charter of Henry II in favor of the London weavers.

Henry, by the grace of God, King of England, Duke of Normandy and Aquitaine, Count of Anjou, to the bishops, justiciars, sheriffs, barons, and all his servants and liegemen of London, greeting. Know that I have granted to the weavers of London to have their guild in London with all the liberties and customs which they had in the time of King Henry, my grandfather. Let no one carry on this occupation unless by their permission, and unless he belongs to their guild, within the city or in Southwark or in the other places pertaining to London, other than those who were wont to do so in the time of King Henry, my grandfather. Wherefore I will, and firmly order that they shall everywhere legally carry on their business, and that they shall have all the aforesaid things as well and peacefully and freely and honorably and entirely as ever they had them in the time of King Henry, my grandfather; provided always that for this privilege they pay me each year 2 marks of gold at Michaelmas. And I forbid anyone to do them injury or insult in respect of this on pain of 10 pounds forfeiture. Witness: Thomas of Canterbury; Warin Fitz Gerold. At Winchester.

58. The Persecution of Jews

Jews settled in England shortly after the Norman Conquest; according to the chroniclers they were to be found in the island as early as the eighth century. As non-Christians they were discriminated against and at times actively persecuted; they lacked the rights of citizens and lived

---

SOURCES: Henry T. Riley (ed.), *Munimenta Gildhallae Londonensis* (London: Rolls Series, 1859–62), *II*, 33, in *English Historical Documents, 1042–1189*, David C. Douglas and George W. Greenaway (eds.), David C. Douglas (gen. ed.),(London: Eyre and Spottiswoode, 1953; New York: Oxford University Press), 947–48.

Henry T. Riley (trans.), *Annals of Roger de Hoveden* (London: G. Bell and Sons, 1853), *II*, 137–38.

segregated in the towns. Unlike the Christians they were not forbidden to lend money at interest, and consequently came to occupy a prominent position in the financial affairs of the realm. For this reason they were afforded some protection by the crown. However, the resentment of debtors combined with religious bigotry ultimately led to their banishment from the kingdom under Edward I. Anti-Jewish feeling was particularly intense in the early months of Richard I's reign (1189–1199); it was then that the Jews of York suffered the attack described below, which is narrated by the chronicler Roger of Hoveden.

In the same month of March, on the seventeenth day before the calends of April,[49] being the sixth day before Palm Sunday, the Jews of the city of York, in number five hundred men, besides women and children, shut themselves up in the tower of York, with the consent and sanction of the keeper of the tower, and of the sheriff, in consequence of their dread of the Christians; but when the said sheriff and the constable sought to regain possession of it, the Jews refused to deliver it up. In consequence of this the people of the city, and the strangers who had come within the jurisdiction thereof, at the exhortation of the sheriff and the constable with one consent made an attack upon the Jews.

After they had made assaults upon the tower day and night, the Jews offered the people a large sum of money to allow them to depart with their lives, but this the others refused to receive. Upon this, one skilled in their laws arose and said: "Men of Israel, listen to my advice. It is better that we should kill one another, than fall into the hands of the enemies of our law." Accordingly, all the Jews, both men as well as women, gave their assent to his advice, and each master of a family, beginning with the chief persons of his household, with a sharp knife first cut the throats of his wife and sons and daughters, and then of all his servants, and lastly his own. Some of them also threw their slain over the walls among the people; while others shut up their slain in the King's house and burned them, as well as the King's houses. Those who had slain the others were afterwards killed by the people. In the meantime, some of the Christians set fire to the Jews' houses and plundered them; and thus all the Jews in the city of York were destroyed, and all acknowledgments of debts due to them were burnt.

[49] March 17, as reckoned by the Roman calendar.

SECTION I

The Course of Religion

59. The Coming of the Cistercians

New impetus was given to the monastic movement in England by the advent of the Cistercians in the early twelfth century. This new order, originating in Citeaux, France, imposed exceptionally rigorous standards of conduct and religious practice upon its members, and was noted for its asceticism. The account given below is from a manuscript of Fountains Abbey, one of the great Cistercian monasteries established in England.

During the reign in England of the illustrious King Henry, son of William, nicknamed the Bastard, the land was kept in peace before his face. Churches were erected in many places, monasteries were built, and in the absence of wars religion spread. There flourished in that time of happy memory, blessed Bernard, Abbot of Clairvaux, a man of splendor, a strenuous worker at the business of God, one who was notable in his holiness and famous in his teaching, being glorious also in the miracles which he wrought. His greatest care was for divine worship and for the salvation of souls. He strove, moreover, to increase his community, giving glory to God with his lips and himself being glorified in His name. He was the spiritual father of many monks, and he caused not a few monasteries to be built. He sent the soldiers of his army to conquer distant lands, and thus he won notable victories over the old enemy, seizing plunder which he could restore to his King. This man, then, moved by divine inspiration, desired to plan an offshoot of good hope from the noble vine of Clairvaux. He therefore sent a chapter of monks to England, hoping to obtain fruit there as well as in so many other lands. There still exists the letter which he wrote to the King, saying that he wished to seize plunder for his heavenly Lord out of the lands of the King, adding that he had sent valiant soldiers from his army who would not be slow to capture what they wanted. In this manner he concluded his entreaty. And this is what was done. His men were received with honor by the

SOURCE: John R. Walbran (ed.), *Memorials of the Abbey of St. Mary of Fountains* (Durham: Surtees Society, 1863), *I*, 3–5, in *English Historical Documents, 1042–1189*, David C. Douglas and George W. Greenaway (eds.), David C. Douglas (gen. ed.), (London: Eyre and Spottiswoode, 1953; New York: Oxford University Press), 692.

King and the kingdom, and they established new fortifications in the province of York. They constructed the abbey which is called Rievaulx, which was the first plantation of the Cistercian Order in the province of Yorkshire. Those who were sent were holy men, being monks who glorified God in the practice of poverty. They dwelt in peace with all men, although they warred with their own bodies and with the old enemy. They showed forth the discipline of Clairvaux whence they came, and by works of piety they spread the sweet savor of their mother-abbey, as it were, a strong perfume from their own house. The story spread everywhere that men of outstanding holiness and perfect religion had come from a far land; that they had converse with angels in their dwelling; and that by their virtues they had glorified the monastic name. Many therefore were moved to emulate them by joining this company whose hearts had been touched by God. Thus very soon they grew into a great company. . . .

60. The Franciscans

The friars first came to England in 1220 when the Dominicans, founded in Spain by the learned St. Dominic a few years earlier, reached the island. The Franciscans came four years later. The friars have been described as "men who had abandoned all that was enticing in life to imitate the apostles, to convert the sinner and unbeliever, to arouse the slumbering moral sense of mankind, to instruct the ignorant, to offer salvation to all."[50] The strict rule and selfless aims of the Franciscans were thus set forth by their founder, St. Francis of Assisi:

Those brothers to whom God has given the ability to labor shall labor faithfully and devoutly; in such way that idleness, the enemy of the soul, being excluded, they may not extinguish the spirit of holy prayer and devotion, to which other temporal things should be subservient. As a reward, moreover, for their labor, they may receive for themselves and their brothers the necessaries of life, but not coin or money; and this humbly, as becomes servants of God and the followers of most holy

---

SOURCES: S. Franco and H. Dalmazzo (eds.), *Bullarum Diplomaticum et Privilegiorum Pontificum* (Turin: A. Tomassetti, 1857–72), *III*, 394, in Ernest F. Henderson, *Select Historical Documents of the Middle Ages* (London: G. Bell and Sons, 1896), 346.

J. S. Brewer and Richard Howlett (eds.), *Monumenta Franciscana* (London: Rolls Series, 1858–82), *I*, 7–8, 17–18, in Margaret A. Hennings, *England under Henry III* (London: Longmans, Green, 1924), 228–29.

---

[50] H. C. Lea, *History of the Inquisition of the Middle Ages* (New York, 1888), I, 266.

poverty. The brothers shall appropriate nothing to themselves, neither a house nor a place nor anything; but as pilgrims and strangers in this world, in poverty and humility serving God, they shall go confidently seeking for alms. Nor need they be ashamed, for the Lord made himself poor for us in this world. This is that height of most lofty poverty, which has constituted you my most beloved brothers heirs and kings of the kingdom of heaven . . .

Thomas of Eccleston, a thirteenth-century Franciscan friar, tells us of the first members of his order to come to England.

These nine having then been charitably conveyed across to England, and cordially provided for in their necessities by the nuns of Fécamp,[51] on arriving at Canterbury sojourned for two days at the priory of the Holy Trinity. Then four of them at once set off for London. . . . The other five went to the Priests' Hospice, where they remained until they found for themselves a dwelling. But very shortly after their arrival they were given a small chamber at the back of a school-house, where from day to day they remained almost continuously shut up. But when the schoolboys had gone home in the evening, the brethren went into the school-house and there made a fire and sat by it. And sometimes at the evening conference they would put on the fire a small pot in which were the dregs of beer, and they would dip a cup into the pot and drink in turn. . . . One who merited to be a companion and participator in this unblemished simplicity and holy poverty has testified that at times the beer was so thick that when the pot was to be put on the fire they had to put in it water, and so they drank rejoicing.

The devout and simple life of the Franciscans made a marked impression on thirteenth-century Englishmen, and the order became a prominent feature of the religious life of the country. Here Thomas of Eccleston goes on to describe the early stages of the movement in England.

After this, through the growing number of friars and their proved holiness, the devotion of the faithful increased so that they took care to supply adequate places for them. And so a certain site in Canterbury was given them, and Alexander, master of the Priests' Hospital, built them a chapel good enough for a time; and because the friars did not wish to own any property at all it was made the property of the community of the city . . .

Robert le Mercer first received the friars in Oxford, and assigned them a house where there entered the order many honorable bachelors[52] and

[51] Located on the French coast, northeast of Le Havre.
[52] Knights of low rank.

nobles. Afterwards they hired from Richard le Muliner a house in which they are now. Within a year he bestowed the site and the house upon the community of the town for the use of the friars. That site, however, was small and too narrow. In Canterbury the burgesses of the town first received the friars, assigning them an ancient synagogue next to the prison. But since the neighborhood of the prison was intolerable to the friars, for the prisoners and the friars had the same entrance, the lord King gave them ten marks to pay a rent which would suffice his exchequer for the rent of the site, and so the friars built a chapel so very meager that one carpenter made it in a single day . . .

61. Bishop Grosseteste's Criticism of the Clergy, 1244

The teachings of the church were accepted with very little questioning in England before the fourteenth century. The conduct of the clergy, however, was occasionally criticized. Robert Grosseteste, a noted scholar who became Bishop of Lincoln, called attention to certain clerical defects in this letter to the archdeacons of his diocese in 1244.

From a trustworthy source we have heard that very many priests in your archdeaconry, "neither fearing God nor reverencing men," either do not say or corruptly say the canonical hours, and what they do say is said without any devotion or mark of devotion, nay rather they speak with the manifest signs of an undevout mind. Nor do they observe an hour which would be more convenient for their parishioners for hearing divine service, but one which agrees better with their wanton sloth. Moreover they have their concubines. . . . Also, as we have heard, the clergy hold games which they call "miracles," and other games which they call "the Introduction of May," or "Autumn," and the lay folk "scotales." This can in no way be hid from you, should you prudently and diligently make inquiries thereupon. But there are certain rectors and vicars and priests who are not only too proud to hear the preaching of the friars of either order, but they maliciously prevent the people, as far as they can, from hearing the preachers or confessing to them. Also, it is said, they admit preachers in search of gain to preach, who preach only the sort of things most suitable for extracting money, though we do not give leave to preach to anyone in search of gain; we only grant that their business may be simply explained through the parish priests. . . .

SOURCE: Robert Grosseteste, *Epistolae*, Henry R. Luard (ed.), (London: Rolls Series, 1861), 317–18, in Margaret A. Hennings, *England under Henry III* (London: Longmans, Green, 1924), 224–25.

62. Faith, Reason and Science

The Middle Ages is often described as an age of faith, and frequently the connotation is one of blind faith. Yet much of the intellectual energy of the "High" Middle Ages (c. 1000 to c. 1300) was devoted to reconciling or integrating orthodox Christian belief with the demands of reason. As the centralized, papal Church became more influential, it defined and enforced its dogmas more rigorously. To deny or question seriously major tenets was to risk the penalties of heresy, and thus the intellectual was constrained to operate within a framework. But in many respects it was a spacious framework, affording considerable scope for acute argumentation, which generation after generation of scholastics (or schoolmen) pursued. The three selections given below are taken from the writings of St. Anselm of Canterbury, John of Salisbury and Roger Bacon.

St. Anselm of Canterbury – Anselm, who was Archbishop of Canterbury, wrote around 1077 a work entitled *Proslogion* (An Address). His attitude is suggested by the alternative title: *Faith in Search of Understanding*. For him reason operates within the context of faith: men must believe in order to understand.

O Lord, in my hunger I began to seek Thee; I beseech Thee, let me not, still fasting, fall short of Thee. Famished I have approached Thee; let me not draw back unfed. Poor as I am, I have come to the wealthy, miserable to the merciful; let me not go back empty and despised. And if "before I eat I sigh,"[53] after my sighs give me something to eat. O Lord, I am bent over and can only look downward; raise me up so that I can reach upward. "My iniquities," which "have gone over my head," cover me altogether, "and as a heavy burden" weigh me down.[54] Rescue me, take away my burden, lest their "pit shut her mouth upon me."[55] Let me receive Thy light, even from afar, even from the depths. Teach me to

---

SOURCES: Anselm, "Proslogion," *The Library of Christian Classics, Vol. X: A Scholastic Miscellany: Anselm to Ockham*, Eugene R. Fairweather (ed.), (Philadelphia: The Westminster Press, 1956; London: Student Christian Movement Press), 72–73.

Daniel D. McGarry (ed.), *The Metalogicon of John of Salisbury* (Berkeley: University of California Press, 1955), 222–23.

Robert B. Burke (ed.), *The Opus Majus of Roger Bacon* (Philadelphia: University of Pennsylvania Press, 1928), *II*, 583–85.

---

[53] Job 3: 24.
[54] Psalms 37: 5.
[55] Psalms 68: 16.

seek Thee, and when I seek Thee show Thyself to me, for I cannot seek Thee unless Thou teach me, or find Thee unless Thou show me Thyself. Let me seek Thee in my desire, let me desire Thee in my seeking. Let me find Thee by loving Thee, let me love Thee when I find Thee.

I acknowledge, O Lord, with thanksgiving, that Thou hast created this Thy image in me, so that, remembering Thee, I may think of Thee, may love Thee. But this image is so effaced and worn away by my faults, it is so obscured by the smoke of my sins, that it cannot do what it was made to do, unless Thou renew and reform it. I am not trying, O Lord, to penetrate Thy loftiness, for I cannot begin to match my understanding with it, but I desire in some measure to understand Thy truth, which my heart believes and loves. For I do not seek to understand in order to believe, but I believe in order to understand. For this too I believe, that "unless I believe, I shall not understand."[56]

John of Salisbury – John of Salisbury (d. 1180), regarded as one of the most learned men of his age, who from obscure English birth rose to be Bishop of Chartres, completed in 1159 a work called the *Metalogicon.* It was written as a defense of the studies of the trivium, the basic medieval curriculum made up of grammar, rhetoric and logic. He entitled one of his chapters (Book IV, Chapter 13) "The difference between 'science' and 'wisdom,' and what is 'faith.'" In the previous chapter he acknowledges that scientific knowledge originates from sensation: "Since sensation gives birth to imagination, and these two to opinion, and opinion to prudence, which grows to the maturity of scientific knowledge, it is evident that sensation is the progenitor of science." He then goes on:

In view of the aforesaid, our forefathers used the words "prudence" and "science" with reference to temporal sensible things, but reserved the terms "understanding" and "wisdom" for knowledge of spiritual things. Thus it is customary to speak of "science" relative to human things, but of "wisdom" with regard to divine things. Science is so dependent on sensation that we would have no science concerning things we know by our senses, if these things were not subject to sense perception. This is clear from Aristotle. Despite what I have said above, opinion can be reliable. Such is our opinion that after the night has run its course, the sun will return. But since human affairs are transitory, only rarely can we be sure that our opinion about them is correct. If, nevertheless, we posit as a certainty something that is not in all respects certain, then we approach the domain of faith, which Aristotle defines as "exceedingly strong

[56] Cf. Isaiah 7: 9.

opinion."[57] Faith is, indeed, most necessary in human affairs, as well as in religion. Without faith, no contracts could be concluded, nor could any business be transacted. And without faith, where would be the basis for the divine reward of human merit? As it is, that faith which embraces the truths of religion deserves reward. Such faith is, according to the Apostle, "a substantiation of things to be hoped for, a testimonial to things that appear not."[58] Faith is intermediate between opinion and science. Although it strongly affirms the certainty of something, it has not arrived at this certainty by science. Master Hugh[59] says: "Faith is a voluntary certitude concerning something that is not present, a certitude which is greater than opinion, but which falls short of science."[60] Here, by the way, the word "science" is used in an extended sense, as including the comprehension of divine things.

Roger Bacon – Roger Bacon (1214?–1294), a Franciscan, opposed the abuses found in the syllogistic form of deductive reasoning employed by the schoolmen; this and his scientific investigations caused him to be regarded with suspicion, and even to be censured by his Order. He made noteworthy scientific contributions, particularly in optics. In his *Opus Majus* (1267) he emphasizes the importance of the experimental method. But the reader will note that he does not rule out "the grace of faith."

Having laid down fundamental principles of the wisdom of the Latins so far as they are found in language, mathematics and optics, I now wish to unfold the principles of experimental science, since without experience nothing can be sufficiently known. For there are two modes of acquiring knowledge, namely, by reasoning and experience. Reasoning draws a conclusion and makes us grant the conclusion, but does not make the conclusion certain, nor does it remove doubt so that the mind may rest on the intuition of truth, unless the mind discovers it by the path of experience; since many have the arguments relating to what can be known, but because they lack experience they neglect the arguments, and neither avoid what is harmful nor follow what is good. For if a man who has never seen fire should prove by adequate reasoning that fire burns and injures things and destroys them, his mind would not be satisfied thereby, nor would he avoid fire, until he placed his hand or some combustible substance in the fire, so that he might prove by experience that which reasoning taught. But when he has had actual experience of

57 *Topica*, iv, 5, 18, 126b.
58 Hebrews 11: 1.
59 Hugh of St. Victor (c. 1097–1141), a French mystical theologian.
60 *Summa Sententiarum*, I, 1.

combustion his mind is made certain and rests in the full light of truth. Therefore reasoning does not suffice, but experience does.

This is also evident in mathematics, where proof is most convincing. But the mind of one who has the most convincing proof in regard to the equilateral triangle will never cleave to the conclusion without experience, nor will he heed it, but will disregard it until experience is offered him by the intersection of two circles, from either intersection of which two lines may be drawn to the extremities of the given line; but then the man accepts the conclusion without any question. Aristotle's statement, then, that proof is reasoning that causes us to know is to be understood with the proviso that the proof is accompanied by its appropriate experience, and is not to be understood of the bare proof. His statement also in the first book of the Metaphysics that those who understand the reason and the cause are wiser than those who have empiric knowledge of a fact, is spoken of such as know only the bare truth without the cause. But I am here speaking of the man who knows the reason and the cause through experience. These men are perfect in their wisdom, as Aristotle maintains in the sixth book of the Ethics, whose simple statements must be accepted as if they offered proof, as he states in the same place.

He therefore who wishes to rejoice without doubt in regard to the truths underlying phenomena must know how to devote himself to experiment. For authors write many statements, and people believe them through reasoning which they formulate without experience. Their reasoning is wholly false. For it is generally believed that the diamond cannot be broken except by goat's blood, and philosophers and theologians misuse this idea. But fracture by means of blood of this kind has never been verified, although the effort has been made; and without that blood it can be broken easily. For I have seen this with my own eyes, and this is necessary, because gems cannot be carved except by fragments of this stone . . . Therefore all things must be verified by experience.

But experience is of two kinds: one is gained through our external senses, and in this way we gain our experience of those things that are in the heavens by instruments made for this purpose, and of those things here below by means attested by our vision. Things that do not belong in our part of the world we know through other scientists who have had experience of them. . . . This experience is both human and philosophical, as far as man can act in accordance with the grace given him; but this experience does not suffice him, because it does not give full attestation in regard to things corporeal owing to its difficulty, and does not touch at all on things spiritual. It is necessary, therefore, that the intellect of man should be otherwise aided, and for this reason the holy patriarchs and prophets, who first gave sciences to the world, received illumination within and were not dependent on sense alone. The same is true of many believers since the time of Christ. For the grace of faith illuminates greatly, as also do divine inspirations, not only in things spiritual, but in things corporeal and in the science of philosophy; as Ptolemy states in the *Centilogium*, namely, that there are two roads by which we arrive at the knowledge of facts, one through the experience of philosophy, the other through divine inspiration, which is far the better way, as he says.

Part 3

THE LATER MIDDLE AGES

1272-1485

SECTION A

Edward I: Parliamentary Developments

63. The Summoning of Representatives

The thirteenth century saw increasing use made of the representative principle. Kings sought to discuss problems of government not only with the great barons and ecclesiastical dignitaries who would be found in the council but also with representatives of the lesser landed interest (knights of the shire) and of the towns (citizens and burgesses). The election of such men, in accordance with a royal order as given below, became a recognized political activity. Note that those elected are to have full powers to act for those whom they represent; they were to participate in the so-called Model Parliament of 1295.

The King to the Sheriff of Northampton, greeting. Whereas we wish to have a conference and discussion with the earls, barons, and other nobles of our realm concerning the provision of remedies for the dangers that in these days threaten the same kingdom – on which account we have ordered them to come to us at Westminster on the Sunday next after the feast of St. Martin[1] in the coming winter, there to consider, ordain, and do whatever the avoidance of such dangers may demand – we command and firmly enjoin you that without delay you cause two knights, of the more discreet and more capable of labor, to be elected from the aforesaid

---

SOURCE: F. Palgrave, *Parliamentary Writs* (London: Record Commission, 1827–34), *I*, 28–30, in Carl Stephenson and Frederick G. Marcham, *Sources of English Constitutional History* (New York: Harper and Row, 1937), 159–60.

---

[1] November 11.

county, and two citizens from each city of the aforesaid county, and two burgesses from each borough, and that you have them come to us on the day and at the place aforesaid; so that the said knights shall then and there have full and sufficient authority on behalf of themselves and the community of the county aforesaid, and the said citizens and burgesses on behalf of themselves and the respective communities of the cities and boroughs aforesaid, to do whatever in the aforesaid matters may be ordained by common counsel; and so that, through default of such authority, the aforesaid business shall by no means remain unfinished. And you are there to have the names of the knights, citizens, and burgesses, together with this writ. By witness of the King, at Canterbury, October 3.

64. The Confirmation of the Charters, 1297

Of fundamental importance in the growth of Parliament as a representative body is the concept that consent to extraordinary taxes (that is taxes apart from those recognized as rightfully belonging to the king, such as feudal dues) should be obtained not only from the feudal and ecclesiastical elements in the royal council, but also from representatives of the nation at large. The earliest precedent for this is to be found in a document, dated 1297, by which Edward I confirmed Magna Carta and the Small, or Forest, Charter. There we find this passage:

And whereas some people of our kingdom are fearful that the aids and taxes, which by their liberality and good will they have heretofore paid to us for the sake of our wars and other needs, shall, despite the nature of the grants, be turned into a servile obligation for them and their heirs because these [payments] may in a future time be found in the rolls . . . [therefore] we have granted, for us and our heirs, that, on account of anything that has been done or that can be found from a roll or in some other way, we will not make into a precedent for the future any such aids, taxes, or prises.[2] And for us and our heirs we have also granted to the archbishops, bishops, abbots, priors, and other folk of Holy Church, and to the earls and barons and the whole community of the land, that on no account will we henceforth take from our kingdom such aids, taxes, and prises, except by the common consent of the whole kingdom

---

SOURCE: William Stubbs, *Select Charters and Other Illustrations of English Constitutional History* (Oxford: The Clarendon Press, 1895), 495, in Carl Stephenson and Frederick G. Marcham, *Sources of English Constitutional History* (New York: Harper and Row, 1937), 164–65.

---

[2] Royal tolls on merchandise.

and for the common benefit of the same kingdom, saving the ancient aids and prises due and accustomed.

SECTION B

Edward I: Government by Statute

65. The Statute of Gloucester, 1278

By legislative measures made in Parliament, and called statutes, Edward I strove to provide permanent guidelines for various policies. Among other things, he undertook to limit the privileges and exemptions of the feudal lords, whom he recognized to be in competition with the royal authority. Shortly after becoming King he sent itinerant justices to conduct an investigation of this situation, with a view to restricting the rights of the holders of private courts to those held by long usage or definite grant. The purpose, as well as the procedure involved, was set forth in the Statute of Gloucester.

And the sheriffs shall cause it to be commonly proclaimed throughout their bailiwicks – that is to say, in cities, boroughs, market towns and elsewhere – that all those who claim to have any franchises by charters of the King's predecessors, kings of England, or in other manner, shall come before the King or before the justices in eyre[3] at a certain day and place to show what sort of franchises they claim to have, and by what warrant. And the sheriffs themselves shall then be there in their proper persons, with their bailiffs and officers, to certify the King upon the aforesaid franchises and other matters touching the same. And this proclamation before the King shall contain warning of three weeks. . . . And if they that claim to have such franchises come not at the day aforesaid, then the franchises shall be taken into the King's hand by the sheriff of the place, in name of distress,[4] so that they shall not use such franchises until they come to receive justice. . . .

SOURCE: *Statutes of the Realm* (London: Her Majesty's Stationery Office, 1810–28), *I*, 45–46.

---

[3] Justices on circuit.

[4] The act of distraint, involving seizure and withholding in satisfaction of a demand or claim.

66. The Statute of Westminster *(Quia Emptores)*, 1290

Another statute, known from the opening words of the Latin text as *Quia Emptores*, in effect prohibited the practice of subinfeudation. It was initiated by the greater landholders; no elected representatives participated in framing this measure. On its military side, feudalism had long been in decline; by this statute a halt was made in the creation of new rungs on the feudal ladder. The ultimate effect was to bring more landholders into closer relationship with the crown, and thus to increase its direct influence.

Forasmuch as purchasers of lands and tenements of the fees[5] of great men and others have many times heretofore entered into their fees, to the prejudice of the lords, to which purchasers the freeholders of such great men and others have sold their lands and tenements to be holden in fee to them and their heirs of their feoffors,[6] and not of the chief lords of the fees, whereby the same chief lords have many times lost their escheats, marriages and wardships of lands and tenements belonging to their fees; which thing seemed very hard and extreme unto those great men and other lords, and moreover in this case manifest disinheritance:

Our lord the King, in his Parliament at Westminster after Easter, the eighteenth year of his reign, . . . at the instance of the great men of the realm, granted, provided and ordained that from henceforth it shall be lawful to every freeman to sell at his own pleasure his lands and tenements, or part of them; so that the feoffee[7] shall hold the same lands or tenements of the same chief lord and by the same services and customs as his feoffor held before. . . .

67. The Statute of Mortmain, 1279

To check the flow of real estate into the hands of the church the Statute of Mortmain was passed. The word *mortmain* means, literally, *dead hand,* and refers to the unbreakable grip the church secured on land, once it came into its possession. By the very nature of the church, as a corporation, certain feudal rights (relief, marriage, wardship) could

SOURCES: *Statutes of the Realm* (London: Her Majesty's Stationery Office, 1810–28), *I*, 106.

Statutes of the Realm (London: Her Majesty's Stationery Office, 1810–28), *I*, 51.

[5] Fiefs.
[6] Held in absolute possession by them and their heirs of those conferring the fiefs.
[7] One put in possession of a fief.

not be exercised over such land. The measure stemmed from the concern of the king and the greater barons over the loss of such valuable perquisites.

The King to his justices of the bench, greeting:

Where of late it was provided that religious men[8] should not enter into the fees of any without licence and will of the chief lords of whom such fees be holden immediately; and notwithstanding such religious men have since entered as well into their own fees as into the fees of other men, appropriating and buying them and sometimes receiving them of the gift of others, whereby the services that are due of such fees, and which at the beginning were provided for the defense of the realm, are wrongfully withdrawn and the chief lords do lose their escheats of the same:

We, therefore, to the profit of our realm, intending to provide convenient remedy, by the advice of our prelates, earls, barons and other our subjects being of our council, have provided, established and ordained that no person, religious or other, whatsoever he be, presume to buy or sell, or under the color of gift or lease or by reason of any other title, whatsoever it be, to receive of any man, or by any other craft or contrivance to appropriate to himself, any lands or tenements under pain of forfeiture of the same whereby such lands or tenements may any wise come into mortmain. . . .

SECTION C

Edward I and the Control of Britain

68. The Conquest of Wales

Under Edward I, Wales was annexed to England by conquest. Edward began his military campaigns in Wales in 1277; by 1284 he had sufficiently reduced the country to establish a new governmental regime, set forth in the Statute of Wales. Certain features of the conquest,

---

SOURCE: Henry R. Luard (ed.), *Annales Monastici* (London: Rolls Series, 1864–69), IV, 287–94, in W. D. Robieson, *The Growth of Parliament and the War with Scotland, 1216–1307* (London: G. Bell and Sons, 1914), 77–80.

---

[8] Men of religion, clerics.

including the slaying of Llewellyn, the principal Welsh leader, are described in the Annals of Oseney, a monastic chronicle of the time.

A.D. 1281 – About the festival of the Annunciation of the Blessed Mary, Llewellyn, violating the peace which he had some time before entered into with the King of England, . . . did not shame, with a large band of robbers, to devastate, plunder and burn, in frequent raids, those lands belonging to the King of England and the marchers[9] which lay nearest to him. He even attacked the castles of Flint and Rhuddlan, which the King had begun to build on the borders of Wales to ward off the threatened attacks of the Welsh. When the King . . . heard the news, he sent off a few of his men immediately to check, even a little, the advance of the Welsh, until he himself could take more serious measures. Then, summoning the nobles of the kingdom, he appointed a Parliament to be held at Worcester on the festival of the Nativity of St. John the Baptist.[10]

Meanwhile Roger de Clifford, who was endeavoring to protect the lands lying next his own from the fury of the marauding bands, was captured, mortally wounded, by David and his accomplices, after several of his family had been cruelly put to death. The King, hearing this, decreed in the Parliament above mentioned that all the nobles of the kingdom should meet him with horse and arms in Wales . . . ; and when a large army assembled, he laid waste, ravaged and burned the strongholds, lands and villages of the Prince of Wales, which lay near him. But the Welsh resisted courageously, and one day, when a detachment from the King's army was advancing somewhat carelessly and allowing itself to become too far separated from the main body, suddenly a countless host of Welshmen, bursting forth from hiding places in the woods and marshes, attacked our men, who were relatively few in number. In the struggle were slain the son of Lord William de Valence, nephew of the lord King . . . and several others, the remainder escaping with difficulty.

The King remained in the region of Rhuddlan . . . , and in the meantime the lord John, Archbishop of Canterbury, was sent to Llewellyn at Snowdon to treat for peace with him, or rather to advise and induce him to observe the peace which he had previously made with the King. . . . But his mission was fruitless, for Llewellyn could not be induced to make peace. While the Archbishop delayed for three days in Snowdon, the English nobles, showing more foolishness than courage, secretly entered Snowdon, thinking that by craft they could seize it by their own unaided strength. But the Welsh, forewarned of their approach, advanced in force against them, and joining battle easily prevailed over the small detachment of nobles and put them to flight. . . . When the Archbishop

[9] Officers responsible for the defense of a border.
[10] June 24.

came down from Snowdon without accomplishing his aim, he uttered sentence of excommunication against Llewellyn as a violator of his oath, and a perjurer, and against David, his brother, and all their accomplices and abettors.

About the same time . . . the lord Edmund . . . together with his brother[11] . . . laid an ambush for the said Llewellyn; for, being informed of his movements by spies, the said Edmund gathered together a large and powerful force, and, more by chance than was imagined at that time, fell in with Llewellyn when he had descended from the mountains of Snowdon for some unknown reason and was traversing the lower ground with the few followers who still adhered to him, and put him, and those of his men who were unable to escape, to death by the sword. The head of the Prince, whom he recognized among the slain, was cut off and sent to the lord King. . . .

A.D. 1282 – The King of England, encouraged by the aforesaid victory, and seeing a way open to him for the fulfillment of his desires, lest there should be any impediment to his carrying his wishes into effect, entered in triumph with his men the safe and secret hiding-place of the Welsh, to wit, the province of Snowdon. . . . Then he was able to control, as master, the castles and fortified places, both within Snowdon and without, except a certain castle, called in their tongue Bere. Into this castle David . . . had in vain introduced a garrison, promising to send them speedy assistance, while he himself took refuge in secret and almost inaccessible woods and swamps. The castle itself was surrounded by an impassable marsh, and possessed no entrance except by narrow paths artificially constructed to overcome the natural difficulties of the ground. When the King found this out, he . . . besieged the defenders so straitly that . . . they were compelled to surrender the castle and trust to the clemency of the King, who graciously granted them freedom of life and limb. Then the King, by a lavish distribution of gifts and presents, entered privily into an agreement with some of the natives who knew the hidden ways and secret retreats, and they, not without joy, compelled David to withdraw from his refuge, and surrendered him to the King, who sent him . . . to be imprisoned, along with his wife and son at Rhuddlan.

This took place about the feast of St. Botulf.[12]. . . About Michaelmas the King, summoning the nobles and mayors of the cities to meet him at Salisbury, held a Parliament, and caused David . . . to be brought before him; and after consideration of his magistrates, had him condemned to death, by advice of the magnates. . . .

[11] Edmund and Roger, sons of Roger de Mortimer.
[12] June 17.

69. The Award of Norham, 1292

A dispute with regard to the royal succession in Scotland gave Edward I an opportunity to extend his authority over that kingdom in 1292. Relying on his claim of overlordship over Scotland, Edward appointed a commission of arbitration, which met at Norham and decided in favor of John Balliol. Edward, in turn, received from Balliol homage and fealty for the fief of Scotland.

Therefore, after a diligent discussion of this matter, by common consent the King adjudged the undivided kingdom to John de Balliol, who was descended from the oldest daughter of David, King of Scots. For Robert de Bruce, between whom and the same John de Balliol the question principally lay, to the exclusion of all the others, although one generation nearer, nevertheless was descended from the second daughter of King David. John de Balliol, on the feast of St. Andrew[13] next following, was solemnly crowned in the church of the Canons Regular at Scone, being seated on the royal stone on which Jacob had supported his head when he was going from Beersheba to Dan. After his coronation, going to the King of England, who was celebrating Christmas at Newcastle-on-Tyne, he did homage to him in these words: "Lord Edward, King of England, overlord of Scotland, I, John de Balliol, King of Scotland, acknowledge myself your liegeman for the whole realm of Scotland and for all things which pertain to it and depend upon it; which kingdom of mine I hold, and ought openly and of right to hold, from you and your heirs, kings of England, with life and limbs and earthly honor, against all men who can live and die." And the King of England accepted his homage in this form, the rights of both being saved. . . .

SOURCE: William Rishanger, *Chronica et Annales,* Henry T. Riley (ed.), (London: Rolls Series, 1865), 135–36, in Edward P. Cheyney, *Readings in English History Drawn from the Original Sources* (Boston: Ginn, 1935), 231.

---

[13] November 30.

SECTION D

The Hundred Years' War: Fourteenth Century Phases

70. The Battle of Crécy, 1346

The long conflict between England and France, called in later times the Hundred Years' War, began in 1337. It was two years before a serious offensive against France was mounted; in 1346 the first decisive encounter, the Battle of Crécy, was won by the English. Though smaller in number than the French armies, the English forces were better drilled and disciplined, and they made much greater use of infantrymen, the archers being four times more numerous than the mounted knights. The engagement is important in the history of military tactics; though its implications were not fully grasped for many years, it may be said to have ended the favored position of cavalry in medieval warfare, and thus to have sounded the knell of traditional feudal military arrangements. The French chronicler, Jean Froissart, gives us a detailed account of the battle.

There is no man, unless he had been present, that can imagine or describe truly the confusion of that day, especially the bad management and disorder of the French. . . . The English, who . . . were drawn up in three divisions, and seated on the ground, on seeing their enemies advance, rose up undauntedly and fell into their ranks. The Prince's[14] battalion, whose archers were formed in the manner of a portcullis, and the men-at-arms in the rear, was the first to do so. . . .

You must know that the French troops did not advance in any regular order, and that as soon as their King came in sight of the English his blood began to boil and he cried out to his marshals, "Order the Genoese forward and begin the battle in the name of God and St. Denis." There were about 15,000 Genoese crossbowmen; but they were quite fatigued, having marched on foot that day six leagues, completely armed and carrying their crossbows, and accordingly they told the constable they were not in a condition to do any great thing in battle. The Earl of Alençon hearing this, said, "This is what one gets by employing such scoundrels, who fall off when there is any need for them." During this

---

SOURCE: Sir John Froissart, *The Chronicles of England, France, Spain, etc.* (London: Everyman's Library, J. M. Dent and Sons, 1911), 44–47.

---

[14] The reference is to Edward, Prince of Wales, known as the Black Prince.

time a heavy rain fell, accompanied by thunder and a very terrible eclipse of the sun; . . . shortly afterwards it cleared up, and the sun shone very bright; but the French had it in their faces, and the English on their backs.

When the Genoese were somewhat in order they approached the English and set up a loud shout in order to frighten them; but the English remained quite quiet and did not seem to attend to it. They then set up a second shout, and advanced a little forward; the English never moved. Still they hooted a third time, advancing with their crossbows presented, and began to shoot. The English archers then advanced one step forward, and shot their arrows with such force and quickness that it seemed as if it snowed. When the Genoese felt these arrows, which pierced through their armor, some of them cut the strings of their crossbows, others flung them to the ground, and all turned about and retreated quite discomfited.

The French had a large body of men-at-arms on horseback to support the Genoese, and the King, seeing them thus fall back, cried out, "Kill me those scoundrels, for they stop up our road without any reason." The English continued shooting, and some of their arrows falling among the horsemen drove them upon the Genoese, so that they were in such confusion they could never rally again.

In the English army there were some Cornish and Welsh men on foot, who had armed themselves with large knives. These, advancing through the ranks of the men-at-arms and archers, who made way for them, came upon the French when they were in this danger, and falling upon earls, barons, knights and squires, slew many . . .

71. The Encouragement of Archery

Again at Poitiers (1356), the feudal pattern of warfare was disrupted by the prowess of the English archers, using the longbow. It is not surprising that the English government, through many proclamations and laws, took pains to foster the continuance of skill in archery. Thus, in 1363, a proclamation was issued as follows:

The King to the Lord Lieutenant of Kent, greeting.

Whereas the people of our realm, gentle and simple alike, were wont formerly in their games to practice skill in archery – whence, by the help of God, it is well known that high honor and advantage came unto our realm, and no mean advantage to ourselves in our feats of war – and that now, the said skill in archery having fallen almost wholly into dis-

---

SOURCE: Thomas Rymer, *Foedera* (London: Record Commission, 1816), *III*, pt. ii, 704, in Edward P. Cheyney, *Readings in English History Drawn from the Original Sources* (Boston: Ginn, 1935), 249.

repute, our people give themselves up to the throwing of stones and of wood and of iron; and some to handball and football and hockey; some to coursing and cock fighting; and some to other unseemly sports that be even less useful and manly; whereby our realm — which God forbid — will soon, as it appeareth, be stripped of archers:

We, wishing that a fitting remedy be found in this matter do hereby command you that, in all places in your county, liberties or no liberties, wheresoever you shall see fit, you have proclamation made to this effect: that every man in the same county, if he be able-bodied, shall, upon holidays, make use, in his games, of bows and arrows, or darts, or both, and learn and practice archery.

Moreover, you are to prohibit all and sundry in our name from such stone, wood, and iron throwing; handball, football, or hockey; coursing and cock fighting; or other such idle games which are of no usefulness; under penalty of imprisonment.

By the King, at Westminster, June 1, 1363.

72. The Treaty of Brétigny, 1360

By 1360 English military successes, which had been highlighted by the victories of Crécy and Poitiers and had brought Edward III's forces before the walls of Paris, were rewarded with the Treaty of Brétigny. This marked the zenith of English domination over France in the fourteenth century. The treaty was long and complex, but the following excerpts will give some idea of its major provisions:

Charles, eldest son of the King of France, Regent of the kingdom, Duke of Normandy, and Dauphin of Vienne, to all those who shall see these letters, greeting.

We make known to you that concerning all debates and discussions whatsoever moved and arisen between Monsieur, the King of France, and us, for him and for ourselves and for all those to whom it appertains, on the one part, and the King of England and all those whom it touches on his side, on the other, for the good of the land, it is agreed, the 8th day of May, the year of grace 1360, at Brétigny of Chartres, as follows: first, that the King of England, along with what he holds in Guienne and Gascony, shall have for himself and his heirs, perpetually and for all time, all the possessions that follow, to be held in the manner which the King of France and his sons or any of his ancestors, kings of France, held them;

SOURCE: E. Cosneau, *Les Grands Traités de la Guerre de Cent Ans* (Paris: A. Picard, 1889), 39–60, in Edward P. Cheyney, *Readings in English History Drawn from the Original Sources* (Boston: Ginn, 1935), 247–48.

that is to say, that which is in domain[15] in domain and that which in fief in fief, in the manner explained below.

The city, the castle, and the country of Poitiers and all the land and county of Poitou.[16] . . .

Likewise, the King of England shall have the castle and the city of Calais. . . .

Likewise, it is agreed that the King of England and his heirs shall have and hold all the islands adjacent to the lands, countries, and places above named, together with all the other islands which the said King of England holds at present. . . .

Likewise, it is agreed that the King of France will pay to the King of England 3,000,000 gold crowns, of which two are worth one noble[17] English money. . . .

And as hostages, who shall remain for the King of France, as well those who were taken prisoners at the battle of Poitiers, as others; that is to say, Monsieur Louis, Count of Anjou, Monsieur John, Count of Poitou, sons of the King of France, the Duke of Orleans, brother of the said King, the Duke of Bourbon, the Count of Blois.[18] . . .

Likewise, it is agreed that the King of France and his eldest son, the Regent, for themselves and for their heirs, kings of France, so soon as it can be done, shall withdraw and depart from every alliance which they have with the Scots, and they will promise, so far as they are able, never hereafter, they or their heirs or those who shall be kings of France hereafter, at any time to give to the King or kingdom of Scotland nor to its subjects, present or to come, comfort, aid, or favor, against the said King of England or against his subjects in any way, and that they will not make any other alliances with the said Scots in any time to come, against the said King and kingdom of England.

And likewise, as soon as possible, the King of England and his eldest son will withdraw and depart from all the alliances which they have with the Flemings. . . .

73. Social Effects of the War

The first stages of the Hundred Years' War, besides the notable victories of Crécy and Poitiers, brought to many an Englishman the loot of pil-

SOURCE: Thomas Walsingham, *Historia Anglicana,* Henry T. Riley (ed.), (London: Rolls Series, 1863–64), *II,* 272, in Dorothy Hughes, *Illustrations of Chaucer's England* (London: Longmans, Green, 1918), 164.

15 Demesne, land retained by a lord for his own use.
16 Other territories are mentioned here.
17 An English coin worth 6s 8d, first minted under Edward III.
18 Other names follow.

laged towns, and this doubtless had some effect on English modes and fashions. The chronicler, Thomas Walsingham (d. 1422?) makes reference to this in his *Historia Anglicana* (English History):

In the year of grace 1248 . . . when peace had been secured throughout the whole of England, it seemed to the English that as it were a new sun rising over the land, on account of the abundance of peace, the plenty of all goods, and the glorious victories. For there was no woman of account who did not possess somewhat of the spoils of Calais, Caen, and of other cities across the sea; garments, furs, pillows and household utensils, tablecloths and necklaces, gold and silver cups, linen cloths and sheets, were to be seen scattered throughout England in different houses. Then the ladies of England began to pride themselves upon the apparel of the ladies of France, and while the latter lamented the loss of these things, so the former were rejoicing at having obtained them.

SECTION E

Fourteenth Century Parliamentary Developments

74. Parliamentary Consent to Taxation

Under Edward III (1327–1377) the Commons gained further recognition of the right of consent to taxation, largely owing to financial pressures imposed on the government by the wars with France. In 1340 a statute, after acknowledging certain extraordinary taxes recently granted, goes on to say:

We, willing to provide for the indemnity of the . . . prelates, earls, barons and others of the commonalty, and also of the citizens, burgesses and merchants . . . , will and grant for us and our heirs, to the same prelates, earls, barons and commons, citizens, burgesses and merchants that the same grant which is so chargeable shall not another time be had forth in example, nor fall to their prejudice in time to come, nor that they be from henceforth charged nor grieved to make common aid, or to sustain charge, if it be not by the common assent of the prelates, earls,

---

SOURCE: *Statutes of the Realm* (London: Her Majesty's Stationery Office, 1810–28), *I*, 290, 374.

barons and other great men and commons of our said realm of England, and that in the Parliament . . .

Again, in 1362, a statute dealing in part with custom duties declared against any subsidy (tax levy) on wool without the consent of Parliament.

The King, . . . having regard for the great subsidy that the Commons have granted now in this Parliament of wools, leather and wool-fells,[19] . . . wills and grants that after the said term [of three years] . . . nothing be taken nor demanded of the said Commons but only the ancient custom of half a mark; and that this grant now made or which hath been made in times past shall not be had in example nor charge of the said Commons in time to come; and that the merchants denizens[20] may pass with their wools, as well as foreigners, without being restrained; and that no subsidy nor other charge be set nor granted upon the wools by the merchants or others from henceforth without the assent of the Parliament. . . .

75. Parliament's Use of the Conditional Grant

Parliament early discovered that it could exercise some control over royal policies through its important function of providing supply. A common device was the attachment of conditions to grants, whereby, in accepting the grant, the king bound himself to various courses of action. Such a grant, for example, foreshadowing the modern practice of appropriation, was made in April, 1348.

Nevertheless, provided that the aid now granted by the said Commons be in no manner turned into wool neither by loan, nor by valuation, nor in other manner be levied nor more hastily, than in the form in which it be granted, and that in the meantime the circuits of the justices, as well of the forest as of common pleas and general inquisitions, cease throughout the land; that the aid be levied, and that the subsidy granted of forty shillings on each sack of wool cease at the end of three years, which will be now at Michaelmas next coming, and that henceforth no such grant be made by the merchants, inasmuch as it is only to the grievance

---

SOURCE: *Rotuli Parliamentorum* (London: Her Majesty's Stationery Office, 1776–77), *II*, 200–01, in George B. Adams and H. Morse Stephens, *Select Documents of English Constitutional History* (New York: Macmillan, 1937), 113–14.

---

[19] Skins from which the wool has not been removed.
[20] Aliens admitted to citizenship.

and charge of the Commons, and not of the merchants who buy the wool at so much the less. And also, that henceforth no imposition, tallage, nor charge by loan, nor of any other sort whatsoever, be put by the Privy Council of our lord the King without their grant and assent in Parliament: and also, that two prelates, two lords, and two justices in this present Parliament be assigned to hear and examine all the petitions previously put forward in the last Parliament by the Commons which have not yet been answered; and with them the petitions now set forth, in the presence of four or six of the Commons chosen by them for this special purpose, so that the said petitions be answered reasonably in the present Parliament, and of those which have been previously answered in full, that the answers be in force without change. And also that the merchants who have evilly deceived our lord the King, and have been extortionate toward his people in the matter of the twenty thousand sacks of wool of loan granted by the Commons to our said lord, be put to answer before the justices having power to hear and determine throughout the countries of England, and that no release nor charter of pardon be allowed them. And that the said justices make inquisition of the false money which ruins the people. And that David Bruce, William Douglas and the other chieftains of Scotland be in no manner released neither for ransom nor on parole. And also that our lord the King restore to the Commons the twenty thousand sacks of wool in time past taken from the Commons by loan and that the aid for the marrying of the daughter of our lord the King cease in the meantime. And that there be no Marshalsea[21] in England, save the Marshalsea of our lord the King, or of the guardian of England when our lord the King shall be out of England, upon these conditions above named and not otherwise. And also, provided that the said conditions be entered on the roll of Parliament as a matter of record, so that there can be remedy if anything to the contrary is attempted in time to come. Thus, the said poor Commons, to their very great mischief[22] grant to our lord the King three fifteenths[23] to be levied for three years commencing at Michaelmas next coming; so that each of the three years one fifteenth and no more be levied, at two terms of the year, at Michaelmas and at Easter, in equal portions. And that the said aid be assigned and kept solely for the war of our lord the King and in no manner for the payment of former debts. And also, if, by the grace of God, peace or long truce be made in the meantime, that the fifteenth for the last of the three years be not levied; but of that fifteenth the grant shall

21 The court of the royal household, responsible for disciplining offenders in the household. Its jurisdiction, as exercised when the royal entourage moved about the country, provoked irritation.
22 Misfortune.
23 Tenths and fifteenths were taxes on movable property; the rate became fixed at a tenth from the towns and a fifteenth from the counties.

lose its force completely. And that letters patent of these conditions, and of the manner of this grant be made under the great seal to all the counties of England, without paying anything therefore. And that the said patents make mention of the great necessity of our lord the King, which has arisen since the last Parliament. And also in case of war with Scotland that the aid granted north of the Trent be turned to the conduct of that war and in defence of that part of the country, as before this time has been done.

76. The Beginnings of Impeachment

Parliament sought, without much success, to secure what later would be called ministerial responsibility. One means toward this end was the process of impeachment, by which the House of Commons preferred charges against ministers or other influential persons suspected of wrongdoing, the peers serving as judges. This procedure was first clearly followed in 1376 against Richard Lyons, customs official and merchant, and several others, who were accused of using their offices for illegal purposes. Impeachment was not often employed, however, until the seventeenth century; then it was frequently invoked in Parliament's struggles with the Crown.

Richard Lyons, merchant of London, was impeached and accused by the . . . Commons of many deceptions, extortions, and other crimes committed by him against the King our lord, and against his people, as well during the time when he was in attendance upon the household and council of the King as also during the time when he was farmer[24] of the subsidies and customs of the King; . . . because he has put and procured to be put upon the wool, wool-fells and other merchandises, certain new impositions without the assent of Parliament, and he has levied and collected those impositions largely for his own use and for the use of those about the King, who are of the said covenant, without the oversight or witness of any comptroller, and without his being charged by record or otherwise except at his will, but he alone is sole treasurer and receiver. . . .

Various loans were also made for the use of the King without necessary cause; and especially one loan of twenty thousand marks which was

---

SOURCE: *Rotuli Parliamentorum* (London: Her Majesty's Stationery Office, 1776–77), *II*, 323, in Edward P. Cheyney, *Readings in English History Drawn from the Original Sources* (Boston: Ginn 1935), 280–81.

---

[24] One who undertakes the collection of taxes, etc., paying a fixed sum for the privilege of retaining them.

recently made in London, and this by the advice of the said Richard and other Privy Councillors about the King, who had agreed with the creditors to receive part of the profit and to be partners secretly in the said loan; to which loan the said Richard furnished his own money and afterwards gained by way of usury from the King his lord, of whose Council he had formerly been a member, a great quantity of money to the great damage and deception of the King. . . .

To which the said Richard, being present in Parliament, replied, that as to the loan made to the King of the twenty thousand marks aforesaid, he was entirely free of blame. And further he said that he had gained thereof neither profit nor gain, nor did he furnish anything to the loan aforesaid, in money nor in anything else; and this he was ready to prove by all reasonable means that might be demanded. And as to the said imposition of ten shillings and twelvepence on the sack of wool, etc., . . . he could not clearly excuse himself of having also levied and collected them and of taking thereof a portion, that is to say, twelvepence from each sack of wool, etc. But this he did, he said, at the express command of our lord the King, and at the prayer and with the consent of the merchants . . . And as to the residue of these impositions, he had caused them to be delivered entirely to the receiver of the Chamber[25] of the King and had accounted therefor fully in the said Chamber. And the said Richard was told that for it he ought to produce the warrant by which he had done the said things. But no warrant nor authorization was produced in Parliament under the seal of the King nor otherwise, save only that he said that he had commandment therefor from the King himself and from his Council to do it. And upon this, testimony was given openly in Parliament that our lord the King has said expressly the day before to certain lords here present in Parliament that he did not know how or in what manner he had conferred such an office upon him; and furthermore, he did not recognize him as his officer. And as to the other articles, the said Richard made no answer; but he said that if he had committed offense or done wrong in any wise, he placed himself at the mercy of the King our lord.

Thereupon the said Richard was ordered to prison during the King's pleasure, and to be put to fine and ransom, according to the amount and heinousness of his offense, and that he lose his liberty of the city of London, and that he never hold office of the King, nor enter the council or the palace of the King. . . .

25 The king's private treasury, as opposed to the more public Exchequer.

SECTION F

The Black Death and Its Effects

77. The Black Death

In 1348 and 1349 England, as well as most of western Europe, was ravaged by a virulent epidemic later referred to as the "Black Death," believed by many authorities to have been the bubonic plague. The mortality was very high; it is possible that, by the end of the fourteenth century, this and other visitations of the disease had reduced the population of the country by one half. The social effects of such a catastrophe were naturally far-reaching. The following account is from a contemporary chronicler, Henry Knighton.

Then the dreadful pestilence made its way through the coast land by Southampton, and reached Bristol, and there perished almost the whole strength of the town, as it were surprised by sudden death; for few kept their beds more than two or three days, or even half a day. Then this cruel death spread on all sides, following the course of the sun. And there died at Leicester, in the small parish of St. Leonard more than 380 persons; in the parish of Holy Cross, 400; in the parish of St. Margaret's, Leicester, 700; and so in every parish, a great multitude. . . .

In the same year there was a great murrain[26] of sheep everywhere in the kingdom, so that in one place more than 5000 sheep died in a single pasture; and they rotted so that neither bird nor beast would approach them. There was great cheapness of all things, owing to the general fear of death, since very few people took any account or riches or property of any kind. A horse that was formerly worth 40s could be had for half a mark, a fat ox for 4s, a cow for 12d, a heifer for 6d, a fat wether for 4d, a sheep for 3d, a lamb for 2d, a large pig for 5d, a stone[27] of wool was worth 9d. Sheep and oxen strayed at large through the fields and among the crops, and there were none to drive them off or herd them, but they perished in remote by-ways and hedges in inestimable numbers throughout all districts, because that there was such great scarcity of

---

SOURCE: Joseph R. Lumby (ed.), *Chronicon Henrici Knighton* (London: Rolls Series, 1889–95), *II*, 58–65, in Dorothy Hughes, *Illustrations of Chaucer's England* (London: Longmans, Green, 1918), 145–49.

---

[26] A pestilence affecting livestock.
[27] Commonly fourteen pounds.

servants that none knew what to do. For there was no recollection of such great and terrible mortality since the time of Vortigern, King of the Britons, in whose day, as Bede testifies, the living did not suffice to bury the dead.

In the following autumn a reaper was not to be had for less than 8d, with his food, a mower for less than 10d, with food. Wherefore many crops rotted in the fields for want of men to gather them. But in the year of the pestilence, as has been said above, of other things, there was so great an abundance of all kinds of corn that they were scarcely regarded. . . .

Master Thomas Bradwardine was consecrated by the Pope as Archbishop of Canterbury, and when he returned to England, he came to London, and was dead within two days. He was renowned above all other clerks in Christendom, especially in theology and other liberal sciences. At this time there was everywhere so great a scarcity of priests that many churches were left destitute, without divine service, masses, matins, vespers, or sacraments. A chaplain was scarcely to be had to serve any church for less than £10 or 10 marks, and whereas when there was an abundance of priests before the pestilence a chaplain could be had for 4, 5 or 11 marks, with his board, at this time there was scarcely one willing to accept any vicarage at £20 or 20 marks. Within a little time, however, vast numbers of men whose wives had died in the pestilence flocked to take orders, many of whom were illiterate, and as it were mere laymen, save so far as they could read a little, although without understanding. . . .

In the meantime the King sent notice into all counties of the realm that reapers and other laborers should not receive more than they had been wont, under a penalty defined by statute;[28] and he introduced a statute for this cause. But the laborers were so arrogant and hostile that they paid no heed to the King's mandate, but if anyone wanted to have them he was obliged to give them whatever they asked, and either to lose his fruits and crops, or satisfy their greed and arrogance. But the King levied heavy fines upon abbots, priors, knights of great and less degree, and others great and small throughout the countryside when it became known to him that they did not observe his ordinance, and gave higher wages to their laborers. . . .

Then the King caused many laborers to be arrested, and sent them to prison, many of whom escaped and went away to the forests and woods for a time, and those who were taken were heavily fined. Others swore that they would not take wages higher than had formerly been the custom, and so were set free from prison. The same thing was done in the

[28] The Statute of Laborers; see p. 126.

case of other laborers in the towns. . . . After the pestilence many buildings both great and small in all cities, towns, and boroughs fell into ruins for want of inhabitants, and in the same way many villages and hamlets were depopulated, and there were no houses left in them, all who had lived therein being dead; and it seemed likely that many such hamlets would never again be inhabited. In the following winter there was such dearth of servants for all sorts of labor as it was believed had never been before. For the sheep and cattle strayed in all directions without herdsmen, and all things were left with none to care for them. Thus necessaries became so dear that what had previously been worth 1d was now worth 4d or 5d. Moreover the great men of the land and other lesser lords, who had tenants, remitted the payment of their rents, lest the tenants should go away, on account of the scarcity of servants and the high price of all things – some half their rents, some more, some less, some for one, two, or three years according as they could come to an agreement with them. Similarly, those who had let lands on labor-rents to tenants as is the custom in the case of villeins, were obliged to relieve and remit these services, either excusing them entirely, or taking them on easier terms, in the form of a small rent, lest their houses should be irreparably ruined and the land should remain uncultivated. And all sorts of food became excessively dear.

78. The Statute of Laborers, 1351

The Black Death seriously dislocated the rural economy of the kingdom. There were still as many manors as ever to be cultivated, but the labor force available had been drastically reduced. As one might expect, the remaining laborers, sensing their increased importance, sought wage boosts and other gains. In the interest of the landlords, the government attempted to stabilize the situation by legislation regulating wages and prices, fixing the former at pre-plague levels. The following provisions are taken from the Statute of Laborers.

Whereas late against the malice of servants which were idle, and not willing to serve after the pestilence, without taking excessive wages, it was ordained by our Lord the King, and by assent of the prelates, earls, barons, and other of his Council, that such manner of servants, as well men as women, should be bound to serve, receiving salary and wages, accustomed in places where they ought to serve in the twentieth year of

---

SOURCE: *Statutes of the Realm* (London: Her Majesty's Stationery Office, 1810–28), *I*, 311–12.

the reign of the King that now is,[29] or five or six years before, and that the same servants refusing to serve in such manner should be punished by imprisonment of their bodies, as in the said statute is more plainly contained; whereupon commissions were made to divers people in every county to inquire and punish all them which offend against the same.

And now, forasmuch as it is given the King to understand in this present Parliament, by the petition of the commonalty, that the said servants having no regard to the said ordinance, but to their ease and singular covetise,[30] do withdraw themselves to serve great men and other, unless they have livery[31] and wages to the double or treble of that they were wont to take the said twentieth year, and before, to the great damage of the great men, and impoverishing of all the said commonalty, whereof the said commonalty prayeth remedy: wherefore in the same Parliament by the assent of the said prelates, earls, barons, and other great men of the said commonalty there assembled, to restrain the malice of the said servants, ordained and established the things underwritten:

First. That carters, ploughmen . . . shepherds, swineherds . . . and all other servants shall take liveries and wages, accustomed the said twentieth year, or four years before, so that in the country, where wheat was wont to be given, they shall take for the bushel ten pence, or wheat at the will of the giver, till it be otherwise ordained. And that they be allowed to serve by a whole year, or by other usual terms, and not by the day. And that none pay in the time of sarcling[32] or hay-making but a penny a day. And a mower of meadows for the acre five pence, or by the day five pence. And reapers of corn in the first week of August two pence, and the second three pence, and so till the end of August, and less in the country where less was wont to be given, without meat or drink or other courtesy to be demanded, given, or taken. And that all workmen bring openly in their hands to the merchant towns their instruments, and there shall be hired in a common place and not privy.

Item. That none take for the threshing of a quarter of wheat or rye over 2d ob.[33] and the quarter of barley, beans, peas, and oats 1d ob. if so much were wont to be given. . . . And that the same servants be sworn two times in the year before lords, stewards, bailiffs and constables of every town, to hold and do these ordinances. And that none of them go out of the town, where he dwelleth in the winter, to serve the summer, if he may serve in the same town, taking as before is said. . . . And that those who refuse to make such oath . . . shall be put in the stocks by the

[29] I.e., 1346–1347.
[30] Covetousness.
[31] See p. 175.
[32] Weeding with a hoe.
[33] *Obolus,* half-penny.

said lords, stewards, etc. . . . by three days or more, or sent to the next gaol, there to remain, till they will justify themselves. And that stocks be made in every town by such occasion betwixt this and the Feast of Pentecost.

Item. That carpenters, masons, and tilers, and other workmen of houses shall not take by the day for their work but in manner as they were wont, that is to say: a master carpenter, 3d, and another, 2d. A master free mason 4d and other masons 3d and their servants 1d ob.; tilers 3d and their knaves[34] 1d ob.; plasterers and other workers of mud walls and their knaves, by the same manner without meat or drink 1s from Easter to St. Michael. And from that time less, according to the rate and discretion of the justices which should be thereto assigned. . . .

79. A Complaint about the Laboring Class, 1376

The legislative attempts to freeze conditions of employment, wages and prices were difficult to enforce. With an unprecedented demand for their services, laborers who were denied the advantages they sought in one place might venture forth to find them elsewhere. The static quality of the life of the poorer classes, so marked a feature of the middle ages, was to some extent disrupted. The following petition to Parliament, representing the point of view of lords and employers, gives us a glimpse of the problem in 1376, a few years before the Peasants' Revolt:

To our lord the King and his sage Parliament, the commons show and entreat that, whereas various ordinances and statutes have been made in divers Parliaments for the punishment of laborers and artificers and other servants, the which have continued subtly, by great malice aforethought, to escape the penalty of the said ordinances and statutes; so that as soon as their masters accuse them of mal-service, or wish to pay them for their services according to the form of the statutes, they take flight and depart suddenly out of their employment, and out of their own district, from county to county, hundred to hundred and town to town, in strange places unknown to their masters; so that they know not where to find them to have remedy or suit against them by virtue of the aforesaid statutes. And if such vagrant servants be outlawed at the suit of party,

---

SOURCE: *Rotuli Parliamentorum* (London: Her Majesty's Stationery Office, 1776–77), *II*, 340, in Dorothy Hughes, *Illustrations of Chaucer's England* (London: Longmans, Green, 1918), 162–64.

---

[34] Servant boys.

there is no profit to the suitor, or harm or chastisement for such fugitive servant, because they cannot be found, and never think of returning to the district where they have served in this way. And that above all greater mischief is the receiving of these vagrant laborers and servants when they have thus fled their masters' service, for they are taken into service immediately in fresh places, at such high wages that example and encouragement is given to all servants to go off into fresh districts, and from master to master, as soon as they are displeased about anything. And for fear of such flight, the commons dare not now challenge or offend their servants, but give them whatever they like to ask, in spite of the statutes and ordinances made to the contrary, and this principally for fear of their being received elsewhere as is said. But if such fugitive servants were universally seized throughout the realm, at their coming to offer their services, and put into the stocks, or sent to the nearest gaol, to stay there until they had confessed whence they had come, and from whose service; and if it were known in all districts that such vagrants were to be arrested in this way and imprisoned, and not received into the service of others, as they are, they would have no desire to flee out of their district as they do, to the great impoverishment, ruin, and destruction of the commons, if remedy be not applied as speedily as possible. And be it known to the King and his Parliament that many of the aforesaid wandering laborers have become mendicant beggars, to lead an idle life, and they generally go out of their own district into cities, boroughs, and other good towns to beg, when they are able-bodied, and might well ease the community by living by their labor and service, if they were willing to serve. Many of them become "staff strykers,"[35] and lead an idle life, and commonly they rob poor people in simple villages, by two, three, or four together, and are evilly suffered in their malice! The greater part generally become strong thieves, increasing their robberies and felonies every day on all sides, in destruction of the realm. . . .

[35] Able-bodied beggars.

SECTION G

Fourteenth Century Religion: the Nationalist Challenge

80. Opposition to Foreigners Holding English Benefices, 1343

In the thirteenth and fourteenth centuries Englishmen, for the most part, acknowledged the spiritual authority of the papacy, as was almost universally the case in western Europe. However, growing national feeling brought about some criticism of papal administrative practices; thus there was opposition to papal taxation, appointments, and jurisdiction insofar as they affected England. The filling of English ecclesiastical positions, by papal mandate, with foreigners had long been resented. In 1343 a petition against this practice was presented to the King in Parliament:

Whereas aliens hold so many benefices in this land, and the alms that were wont to be distributed from them are withdrawn, much of the treasure of the land is carried beyond the sea, in maintenance of your enemies,[36] the secrets of the realm are revealed, and your liege clerks in this country have the less advancement; and now lately several cardinals have been made, to two of whom the Pope has granted benefices in this land amounting to 6000 marks a year. . . . And the commons have heard that one of these two cardinals, to wit, the Cardinal of Perigord, is the King's fiercest enemy at the papal court, and the most hostile to his interests. And from year to year the country will be so filled with aliens that it may be a great peril, and scarcely any clerk over here, the son of a great lord or other, will find a benefice wherewith he may be advanced. Wherefore the commons beg a remedy, for they cannot, and will not, endure it longer . . . and that it may please the King to write to the Apostolic See . . . requesting the Pope to suspend these charges and recall what has been done. . . .

Answer – The King is advised of this mischief, and he is willing that remedy and amendment may be ordained, between the great men and commons, if it can be agreed upon; and also he wills and agrees that good

---

SOURCE: *Rotuli Parliamentorum* (London: Her Majesty's Stationery Office, 1776–77), *II*, 143–144, in Dorothy Hughes, *Illustrations of Chaucer's England* (London: Longmans, Green, 1918), 178–79.

---

[36] The papacy was centered at Avignon at this time, and England was at war with France.

letters shall be made to the Pope touching this matter, as well in the name of the King and the great men as of the commons.

81. A Demand for English Cardinals, 1350

The transference of the papal court to Avignon, in southeastern France, in 1308 contributed significantly to antipapal sentiment in England, where the Church was regarded as under the control of French interests. It is not surprising that in 1350 a demand was made for the creation of some English cardinals. According to a chronicler, Geoffrey le Baker:

The King wrote to the Supreme Pontiff entreating that he would advance some clerk of his realm to a cardinal's rank, declaring that he wondered greatly, in that the Roman court had not vouchsafed to receive any Englishman into that sacred order for a long time. . . . The Pope wrote in reply, that the King should make choice among the clerks of his realm of those most fit for this honor, and the Holy Father would willingly consent to his desire touching those chosen, provided that they were approved by the verdict of the cardinals, as is meet for the sought-for dignity, to the honor of God and the Universal Church.

The King chose Master John Bateman, Bishop of Norwich, and Ralph de Stratford, Bishop of London, and presented them by his letters to the papal court; and they awaited the completion of the business at the said court for a long time, but in vain. In the meantime, however, John of Valois[37] . . . presented many of his clerks for promotion by the grace of the Apostolic See, from whom the Pope created twelve cardinals. . . .

82. The Restriction of Papal Jurisdiction

A statute of 1353, subsequently reenacted on various occasions, made it a serious offense for anyone to take cases falling within the jurisdiction of an English law court to a foreign tribunal. These measures were known as statutes of praemunire (from the Latin, "to forewarn") and were primarily aimed at restricting the judicial powers of the papacy.

SOURCES: Geoffrey le Baker, *Chronicon Angliae*, J. A. Giles (ed.), (London: Caxton Society, 1847), 206–07, in Dorothy Hughes, *Illustrations of Chaucer's England* (London: Longmans, Green, 1918), 183–84.

Statutes of the Realm (London: Her Majesty's Stationery Office, 1810–28), *I*, 329.

---

[37] John II, King of France, 1350–1364.

Because it is shown to our lord the King . . . how that divers of the people be and have been drawn out of the realm to answer of things whereof the cognizance pertaineth to the King's court; and also that the judgments given in the same court be impeached[38] in the court of another, in prejudice and disherison of our lord the King, and of his crown, and of all the people of his said realm, and to the undoing and destruction of the common law of the same realm . . . :

It is assented and accorded by our lord the King and the great men and commons . . . that all the people of the King's liegeance of what condition that they be which shall draw any out of the realm in plea, whereof the cognizance pertaineth to the King's court, . . . or which do sue in the court of another to defeat or impeach the judgments given in the King's court, shall have a day . . . to appear before the King and his council, or in his Chancery, or before the King's justices in . . . the one bench or the other, or before other of the King's justices which to the same shall be deputed, to answer in their proper persons to the King of the contempt done in this behalf; and if they come not at the said day . . . , they, their procurators,[39] attorneys, executors, notaries and maintainors shall from that day forth be put out of the King's protection, and their lands, goods and chattels forfeit to the King, and their bodies, wheresoever they may be found, shall be taken and imprisoned and ransomed at the King's will . . .

SECTION H

Fourteenth Century Religion: the Wycliffite Challenge

83. Wycliffe on the Bible in English

John Wycliffe believed the Bible should be made available to the people in their native tongue, a position not generally favored by the Church, which employed the Latin version, or Vulgate, of St. Jerome. Some of Wycliffe's arguments in this matter may be found in his *De Officio Pastoralis* (Concerning the Office of the Pastor), written about 1378.

SOURCE: John Wycliffe, *De Officio Pastoralis,* in Dorothy Hughes, *Illustrations of Chaucer's England* (London: Longmans, Green, 1918), 167.

---

[38] Discredited.
[39] Agents.

The friars and their supporters say that it is heresy to write . . . God's law in English, and make it known to unlearned men. . . .

It seemeth first that the wit[40] of God's law should be taught in that tongue that is more known. . . . Also the worthy realm of France, notwithstanding all hindrance, hath translated the Bible and the Gospels, with other true sentences of doctors out of Latin into French; why should not Englishmen do so? As lords of England have the Bible in French, so it were not against reason that they should have the same sentence in English. . . .

. . . And heretofore friars have taught in England the Paternoster in the English tongue, as men see in the play of York, and in many other districts. Since the Paternoster is part of Matthew's Gospel, as clerks know, why should not all be turned into English? . . . Especially since all Christian men, learned and unlearned, that shall be saved, must always follow Christ, and know his lore and his life. But the common Englishmen know it best in their mother tongue. . . .

84. Wycliffe's Condemnation

John Wycliffe, fourteenth century priest and scholar, was at first essentially a reformer. Troubled by abuses growing out of the contemporary ecclesiastical system, he became in the course of time a heretic, denying established doctrines of the church. The following conclusions attributed to Wycliffe were condemned in London in 1382, as heretical and erroneous.

Heretical conclusions, contrary to the determination of the Church: . . .

I. That the material substance of bread and wine remains after consecration in the sacrament of the altar.

II. That accidents do not remain without a subject in the same sacrament.

III. That Christ is not in the sacrament of the altar inherently, truly, and really, in His own corporal presence.

IV. That if a bishop or priest be in mortal sin he may neither ordain, consecrate, nor baptize.

SOURCE: *Fasciculi Zizaniorum Magistri Johannis Wyclif* (London: Rolls Series, 1858), 277–82, in A. A. Locke, *War and Misrule, 1307–99* (London: G. Bell and Sons, 1913), 75–77.

---

40 Wisdom.

V. That if a man be duly penitent all outward confession is superfluous or useless to him.

VI. To tenaciously affirm that it is not stated in the Gospels that Christ ordained the Mass.

VII. That God should obey the Devil.

VIII. That if the Pope be a worthless and evil man and so a member of the Devil, he has no power over the faithful of Christ granted to him by any, except perchance by Caesar.

IX. That after Urban VI[41] no one is to be received as Pope, but it is necessary for us to live, like the Greeks, under our own laws.

X. To assert that it is contrary to Scripture that ecclesiastical men should have temporal possessions.

Erroneous conclusions contrary to the determination of the Church: . . .

XI. That no prelate should excommunicate anyone unless he first knows he is excommunicated by God.

XII. That if he excommunicates he is thereby a heretic or excommunicate.

XIII. That a prelate excommunicating a clerk who has appealed to the King and the Council of the realm is thereby a traitor to God, King, and realm.

XIV. That those who abstain from preaching or hearing the Word of God or the Gospel preached, on account of the excommunication of men, are themselves excommunicate, and in the day of judgment shall be held to be traitors to God.

XV. To assert that it is lawful to anyone, either deacon or priest, to preach the Word of God without the authority of the Apostolic See, or a Catholic bishop, or some other sufficiently authorized.

XVI. To assert that no one is a civil lord, bishop, or prelate while he is in mortal sin.

XVII. That temporal lords can at their will take away temporal goods from ecclesiastics habitually sinful, or that the public may at their will correct sinful lords.

XVIII. That tithes are pure alms, and that the parishioners may detain them on account of the sins of their curates and confer them at pleasure on others.

XIX. That special prayers restricted to one person by prelates or religious do no more avail the same person, other things being equal, than general prayers.

XX. That the fact of anyone entering into any private religion makes him more unfit and unable to perform God's commandments.

[41] The Pope at the time. His reign (1378–1389) witnessed the beginning of the papal schism.

XXI. That holy men who endow private religions either of possessioners[42] or mendicants had sinned in so endowing.

XXII. That the religious living in private religions are not of the Christian religion.

XXIII. That friars are bound to get their living by the labor of their hands and not by begging.

XXIV. That anyone conferring alms on friars or preaching friars is excommunicate; as is the one who receives.

85. The Lollards

Wycliffe's followers, known as Lollards, were comprised of three principal groups: scholars, associated with him at Oxford, humbler adherents, clerical and lay, and some of the country gentry. The term, *Lollard,* was used in disrespect, and bore some relation to similar words, in Dutch and English, meaning a mumbler or a loafer. The unsympathetic attitude generally held toward them by the clergy is shown in this description of their manners and customs, written about 1382 by the unknown continuer of Knighton's *Compilatio de Eventibus Angliae* (Compilation of English Events):

So they were called by the people disciples of Wycliffe and Wycliffites, and a very great many were thus senselessly beguiled into their sect. The original false Lollards, at the first introduction of this unspeakable sect, wore, for the most part, russet garments, showing outwardly, as it were, the simplicity of their hearts, the more subtly to attract the minds of beholders, and the more securely to set about teaching and sowing their insane doctrine. . . . For their opinions prevailed to such an extent that the half, or the greater part of the people were won over to them, some sincerely, others from shame or fear; since they extolled their adherents as worthy of all praise, immaculate, and conspicuous for their goodness, even although their public and private vices might be well known. . . . On the other hand, those who did not join or favor them, but continued to observe the ancient and sound doctrines of the Church, they declared impious, depraved, malignant and perverse, worthy of all censure, and opponents of the law of God. . . . They asserted that only those turned

---

SOURCE: Joseph R. Lumby (ed.), *Chronicon Henrici Knighton* (London: Rolls Series, 1889–95), *II*, 184–88, in Dorothy Hughes, *Illustrations of Chaucer's England* (London: Longmans, Green, 1918), 197–99.

---

42 Members of religious orders with property, as distinguished from orders supposed to subsist upon alms.

from them who were malignant and confirmed sinners . . . not keeping the law of God which they themselves preached; and constantly in all their sayings they made use of this same ending, always alleging "God's law.". . .

In all their discourses, their doctrine appeared at first to be devout and full of sweetness, but declining towards the end it became abundant in subtle malice and slander. And although newly converted, and but lately imitators of that sect, all their disciples at once adopted in extraordinary fashion the same manner of speech . . . and suddenly changing their natural phrase, both men and women became teachers of the evangelic doctrine. . . .

Thus from the day of their coming into the kingdom there has been violent dissension, because . . . they stirred up son against father, . . . the mother against her son's wife, the servants of a household against their master, and, as it were, every man against his neighbor. . . . Never, probably, since the foundation of the Church, were such suspicion, discord and dissension to be seen. They adopted fittingly enough the name of "Wycliffe's disciples," applied to them everywhere by others; for just as Master John Wycliffe was powerful beyond others in disputation, being held second to none in argument, so these people, even when newly attracted to the sect, became surpassingly eloquent, getting the better of others in all subtleties and wordy encounters, mighty in words and prating. . . . So that what they might not achieve by reasoning, they made up for by quarrelsome violence, with angry shouting and bombastic words. . . .

These Wycliffites . . . incited both men and women to reject the teaching of all others, instructing them on no account to listen to the mendicant friars, whom they dubbed "false preachers"; they constantly plotted against them, . . . and clamorously proclaimed themselves true and evangelical preachers, in that they had the Gospel translated into the English language. And as a result . . . the mendicant friars were looked upon with hatred by many in those days, whereby the Wycliffites, emboldened, strove the more to turn the people's hearts from them, and to hinder them from preaching and asking alms. . . .

86. A Statute against Heretics, 1401

In 1401 Parliament passed an act, at the instance of the bishops, whereby the government undertook to assist the church in maintaining orthodoxy, even to the extent of apprehending heretics and burning at the stake those who stood condemned (hence the common title, "On the Burning of Heretics"). The law specifically refers to the "new sect" of Lollards.

Our sovereign lord the King, . . . by the assent of the estates and other discreet men . . . assembled in Parliament, has granted, established and ordained that no one within the . . . realm or any other dominions subject to his royal Majesty shall presume to preach openly or secretly without first seeking and obtaining the license of the local diocesan, always excepting curates in their own churches, persons who have hitherto been so privileged, and others permitted by canon law; and that henceforth no one either openly or secretly shall preach, hold, teach or instruct, or produce or write any book, contrary to the Catholic faith or the determination of holy Church, nor shall any of the [Lollard] sect hold conventicles[43] anywhere or in any way keep or maintain schools for its wicked doctrines and opinions; and also that henceforth no one shall in any way favor anybody who thus preaches, conducts such or similar conventicles, keeps or maintains such schools, produces or writes such books, or in any such manner teaches, informs or excites the people. . . .

And if any person within the said kingdom and dominions is convicted by sentence before the local diocesan or his commissioners of the said wicked preachings, doctrines, opinions, schools and heretical and erroneous instruction, or any of them, and if he refuses duly to abjure the same . . . , then the sheriff of the county . . . and the mayor and sheriffs or sheriff, or the mayor and bailiffs of the city, town or borough . . . nearest the said diocesan or his said commissioners . . . , shall, after such sentences are proclaimed, receive those persons . . . and shall cause them to be burned before the people in a prominent place, in order that such punishment may strike fear into the minds of others, to the end that no wicked doctrines and heretical and erroneous opinions (against the Catholic faith, the Christian law, and the determination of holy Church), nor their authors and favorers, be sustained . . . or in any way tolerated. . . .

SOURCE: *Statutes of the Realm* (London: Her Majesty's Stationery Office, 1810–28), *II*, 126–28.

---

[43] Unauthorized gatherings for worship.

SECTION I

The Peasants' Revolt

87. John Ball Exhorts the Peasants

There was considerable discontent among Englishmen in the generation following the Black Death, and nowhere was this more evident than among the peasants, who resented the Statute of Laborers and other attempts to maintain the old system of serfdom. Men whom we would call agitators, some of them Lollards, were active in haranguing the lower classes. Best known among them was John Ball, a priest, who inveighed against the iniquity of bondage and had some part in organizing the Peasants' Revolt of 1381. The chronicler Froissart refers to Ball and his message in these words:

John Ball was accustomed to assemble a crowd around him in the marketplace and preach to them. On such occasions he would say, "My good friends, matters cannot go on well in England until all things shall be in common; when there shall be neither vassals nor lords; when the lords shall be no more masters than ourselves. How ill they behave to us! For what reason do they thus hold us in bondage? Are we not all descended from the same parents, Adam and Eve? And what can they show, or what reason can they give, why they should be more masters than ourselves? They are clothed in velvet and rich stuffs, ornamented with ermine and other furs, while we are forced to wear poor clothing. They have wines, spices and fine bread, while we have only rye and the refuse of the straw, and when we drink it must be water. They have handsome seats and manors, while we must brave the wind and rain in our labors in the field; and it is by our labor they have wherewith to support their pomp. We are called slaves, and if we do not perform our service we are beaten, and we have no sovereign to whom we can complain or who would be willing to hear us. Let us go to the King and remonstrate with him; he is young, and from him we may obtain a favorable answer, and if not we must ourselves seek to amend our condition." . . .

SOURCE: Sir John Froissart, *The Chronicles of England, France, Spain, etc.* (London: Everyman's Library, J. M. Dent and Sons, 1911), 207–08.

88. The Peasants Seize London, 1381

In 1381 the peasantry of southeastern England rose in revolt, primarily in an attempt to better their social and economic conditions. Within a short time the rebellion spread as far west as Somerset and as far north as York. The following account describes the attack on London, and subsequent events; it is from the pen of the contemporary chronicler, Henry Knighton.

In the year 1381, the second[44] of the reign of King Richard Second, during the month of May, on Wednesday, the fourth day after the feast of Trinity, that impious band began to assemble from Kent, from Surrey, and from many other surrounding places. Apprentices also, leaving their masters, rushed to join these. And so they gathered on Blackheath,[45] where, forgetting themselves in their multitude, and neither contented with their former cause nor appeased by smaller crimes, they unmercifully planned greater and worse evils and determined not to desist from their wicked undertaking until they should have entirely extirpated the nobles and great men of the kingdom.

So at first they directed their course of iniquity to a certain town of the Archbishop of Canterbury called Maidstone, in which there was a jail of the said archbishop, and in the said jail was a certain John Ball, a chaplain who was considered among the laity to be a very famous preacher; many times in the past he had foolishly spread abroad the word of God, by mixing tares with wheat, too pleasing to the laity and extremely dangerous to the liberty of ecclesiastical law and order, execrably introducing into the church of Christ many errors among the clergy and laymen. For this reason he had been tried as a clerk and convicted in accordance with the law, being seized and assigned to this same jail for his permanent abiding place. On the Wednesday before the feast of the Consecration they came into Surrey to the jail of the King at Marshalsea, where they broke the jail without delay, forcing all imprisoned there to come with them to help them; and whomsoever they met, whether pilgrims or others of whatever condition, they forced to go with them.

On the Friday following the feast of the Consecration they came over

---

SOURCE: Joseph R. Lumby (ed.), *Chronicon Henrici Knighton* (London: Rolls Series, 1889–95), *II*, 131–38, in Edward P. Cheyney, *Readings in English History Drawn from the Original Sources* (Boston: Ginn, 1935), 261–65.

---

44 Actually, the fourth and fifth.
45 An open common on the outskirts of London.

the bridge to London; here no one resisted them, although, as was said, the citizens of London knew of their advance a long time before; and so they directed their way to the Tower[46] where the King was surrounded by a great throng of knights, esquires, and others. It was said that there were in the Tower about one hundred and fifty knights together with one hundred and eighty others, with the mother of the King, the Duchess of Britanny, and many other ladies; and there was present, also, Henry, Earl of Derby, son of John, Duke of Lancaster, who was still a youth; so, too, Simon of Sudbury, Archbishop of Canterbury and Chancellor of England, and Brother Robert de Hales, Prior of the Hospital of England and Treasurer of the King.

John Leg and a certain John, a Minorite,[47] a man active in warlike deeds, skilled in natural sciences, an intimate friend of Lord John, Duke of Lancaster, hastened with three others to the Tower for refuge, intending to hide themselves under the wings of the King. The people had determined to kill the Archbishop and the others above mentioned with him; for this reason they came to this place, and afterwards they fulfilled their vows. The King, however, desired to free the Archbishop and his friends from the jaws of the wolves, so he sent to the people a command to assemble outside the city, at a place called Mile End, in order to speak with the King and to treat with him concerning their designs. The soldiers who were to go forward, consumed with folly, lost heart, and gave up, on the way, their boldness of purpose. Nor did they dare to advance, but, unfortunately, struck as they were by fear, like women, kept themselves within the Tower.

But the King advanced to the assigned place, while many of the wicked mob kept following him. More, however, remained where they were. When the others had come to the King they complained that they had been seriously oppressed by many hardships and that their condition of servitude was unbearable, and that they neither could nor would endure it longer. The King, for the sake of peace, and on account of the violence of the times, yielding to their petition, granted to them a charter with the great seal, to the effect that all men in the kingdom of England should be free and of free condition, and should remain both for themselves and their heirs free from all kinds of servitude and villenage forever. This charter was rejected and decided to be null and void by the King and the great men of the kingdom in the Parliament held at Westminster in the same year, after the feast of St. Michael.

While these things were going on, behold those degenerate sons, who still remained, summoned their father the Archbishop with his above-

46 The Tower of London.

47 Franciscan.

mentioned friends without any force or attack, without sword or arrow, or any other form of compulsion, but only with force of threats and excited outcries, inviting those men to death. But they did not cry out against it for themselves, nor resist, but, as sheep before the shearers, going forth barefooted with uncovered heads, ungirt, they offered themselves freely to an undeserved death, just as if they had deserved this punishment for some murder or theft. And so, alas! before the King returned, seven were killed at Tower Hill, two of them lights of the kingdom, the worthy with the unworthy. John Leg and his three associates were the cause of this irreparable loss. Their heads were fastened on spears and sticks in order that they might be told from the rest. . . .

Whatever representatives of the law they found or whatever men served the kingdom in a judicial capacity, these they slew without delay. . . .

On the following day, which was Saturday, they gathered in Smithfield, where there came to them in the morning the King, who although only a youth in years yet was in wisdom already well versed. Their leader, whose real name was Wat Tyler, approached him; already they were calling him by the other name of Jack Straw. He kept close to the King, addressing him for the rest. He carried in his hand an unsheathed weapon which they call a dagger, and, as if in childish play, kept tossing it from one hand to the other in order that he might seize the opportunity, if the King should refuse his requests, to strike the King suddenly (as was commonly believed); and from this thing the greatest fear arose among those about the King as to what might be the outcome.

They begged from the King that all the warrens,[48] and as well waters as park and wood, should be common to all, so that a poor man as well as a rich should be able freely to hunt animals everywhere in the kingdom — in the streams, in the fish ponds, in the woods, and in the forest; and that he might be free to chase the hare in the fields, and that he might do these things and others like them without objection. When the King hesitated about granting this concession Jack Straw came nearer, and, speaking threatening words, seized with his hand the bridle of the horse of the King very daringly. When John de Walworth, a citizen of London, saw this, thinking that death threatened the King, he seized a sword and pierced Jack Straw in the neck. Seeing this, another soldier, by name Radulf Standyche, pierced his side with another sword. He sank back, slowly letting go with his hands and feet, and then died. A great cry and much mourning arose: "Our leader is slain." When this dead man had been meanly dragged along by the hands and feet into the church of St. Bartholomew, which was near by, many withdrew from the band, and,

[48] Hunting preserves, for small game.

vanishing, betook themselves to flight, to the number it is believed of ten thousand. . . .

89. The Peasants' Gains Revoked

Following the dispersal of the rebels from London, the government revoked the concessions it had been forced to make. This was done through instructions addressed to the sheriffs of all counties, of which the following is an example.

Although lately in the abominable disturbance horribly made by some of our liege subjects who made insurrection against our peace, certain [of] our letters patent were made at the importunate demand of these same insurgents, to the effect that we enfranchised all our liege subjects . . . of certain counties, freeing and quitting them of all bondage and service, and also that we pardoned them for all insurrection made against us . . . granting each and all of them our firm peace; that we willed that they should be free to buy and sell in all cities, boroughs, market towns and other places within the realm; and that no acre of land in the aforesaid counties held in service or bondage should be held at more than 4d the acre. . . . Because, however, the said letters were issued unduly and without mature consideration, we, considering the grant of the aforesaid liberties highly prejudicial . . . , tending manifestly to the disherison[49] of the prelates, lords and magnates and the holy church of England, and to the loss and damage of the state, with the advice of our council we have . . . recalled and annulled the said letters. . . .

SOURCE: Thomas Rymer, *Foedera* (London: Record Commission, 1816), *IV*, 126, in Dorothy Hughes, *Illustrations of Chaucer's England* (London: Longmans, Green, 1918), 236.

---

49 Disinheritance.

SECTION J

The Deposition of Richard II

90. The Statute of Treasons, 1352

The special position and authority of kings was in the fourteenth century given additional insurance by the passage of legislation regarding treason. Hitherto the interpretation of treason had been left to the judges; now it was more clearly defined under the principal headings of compassing the king's death, levying war against him and adhering to his enemies.

. . . Whereas divers opinions have hitherto existed as to what case should be adjudged treason and what not, the King, at the request of the Lords and of the Commons, hath made a declaration in the manner that hereafter followeth, that is to say: when a man doth compass or imagine the death of our lord the King, or of our lady his wife, or of their eldest son and heir; or if a man do violate the King's wife, or the King's eldest unmarried daughter, or the wife of the King's eldest son and heir; or if a man do levy war against our lord the King in his realm, or be adherent to the King's enemies in his realm, giving to them aid and comfort in the realm or elsewhere, and thereof be provably attainted[50] of open deed by people of their condition; or if a man counterfeit the King's great or privy seal, or his money; or if a man bring false money into this realm, . . . knowing the money to be false, to merchandise or make payment in deceit of our said lord the King and of his people; or if a man slay the Chancellor, Treasurer or the King's Justices . . . being in their places doing their offices. . . .

91. The Weapon of Deposition

Despite the strengthened concept of treason, the fourteenth century bore unmistakable evidence that kings might be deposed, before that course was actually taken with regard to Richard II. Edward II had

---

SOURCES: *Statutes of the Realm* (London: Her Majesty's Stationery Office, 1810–28), *I*, 319–20.

Joseph R. Lumby (ed.), *Chronicon Henrici Knighton* (London: Rolls Series, 1889–95), *II*, 219, in George B. Adams and H. Morse Stephens, *Select Documents of English Constitutional History* (New York: Macmillan, 1937), 150.

---

50 Accused.

suffered such a fate in 1327. In 1386 Richard II was warned by the Duke of Gloucester (his uncle) and the Bishop of Ely, in an address purporting to come from the whole Parliament, that he too might be dethroned. Our source is Knighton's Chronicle.

Yet one other thing remains of our message for us to announce to you on the part of your people. For they have it from an old statute,[51] and in fact not very long ago put into force, which is to be regretted, that if the king from any malignant design or foolish contumacy or contempt or wanton wilfulness or in any irregular way should alienate himself from his people, and should not be willing to be governed and regulated by the laws, statutes and laudable ordinances of the realm with the wholesome advice of the lords and peers of the realm, but should headily and wantonly by his own mad designs work out his own private purpose, then it should be lawful for them with the common assent and consent of the people of the realm to depose the king himself from the royal throne and to elevate to the royal throne in his place some near kinsman of the royal line.

92. Charges against Richard II, 1399

In 1399 Richard II was overpowered by his cousin, Henry of Lancaster. Thrown into the Tower of London, he was forced to abdicate in the presence of Parliament. The following day the Lancastrian leaders presented thirty-three charges against him. They can hardly be called such in the usual legal sense, since Richard had already been condemned; rather they are to be looked upon as justifications, in the eyes of the Lancastrians, for his ouster. They included the following:

Item, when the King asked and received many sums by way of loan from many lords and others of the realm, to be repaid at a certain term, notwithstanding that he had promised each individual from whom he received these loans, by his letters patent that he would repay them at the time appointed, he did not fulfill his promise, nor has satisfaction yet been made for the money, whence the creditors are greatly distressed. . . .

Item, whereas the King of England is able to live becomingly upon the issues of his realm and the estates belonging to the crown, without bur-

---

SOURCE: *Rotuli Parliamentorum* (London: Her Majesty's Stationery Office, 1776–77), *III*, 419–20, in Dorothy Hughes, *Illustrations of Chaucer's England* (London: Longmans, Green, 1918), 290–92.

---

[51] There was no such statute; the spokesmen probably had in mind documents relating to Edward II's deposition.

dening his people, when that the realm be not charged with the expenses of war, the same King although during almost the whole of his time there were truces between the realm of England and its enemies, not only gave away the greater part of his patrimony to unworthy persons, but on account of this every year charged his people with so many burdensome grants that they were sorely and excessively oppressed, to the impoverishment of his realm; not applying the money so raised to the common profit and advantage of his realm, but lavishly dissipating it upon his own pomp, display and vain glory. And great sums of money are still owing for provisions for his household, and for his other purchases, although he had wealth and treasure more than any of his predecessors within memory.

Item, being unwilling to protect and preserve the just laws and customs of the realm . . . frequently, from time to time, when the laws were declared and set forth to him by the justices and others of his Council, and he should have done justice to those who sought it according to those laws – he said expressly, with harsh and insolent looks, that his laws were in his own mouth, and sometimes, within his breast; and that he alone could change or establish the laws of his realm. Deceived by which opinion, he would not allow justice to be done to many of his lieges, but compelled numbers of persons to desist from suing common right by threats and fear.

Item, after that certain statutes were established in his Parliament, which were binding until they should be especially repealed by the authority of another Parliament, the King, desiring to enjoy such liberty that no such statutes might restrain him . . . cunningly procured petition to be put forward in Parliament on behalf of the community of the realm, and to be granted him in general – that he might be as free as any of his predecessors; by color of which petition and concession the King frequently caused and commanded many things to be done contrary to such statutes then unrepealed.

Item, although by statute and custom of the realm, upon the summons of Parliament the people of each county ought to be free to choose and depute knights for the county to be present in Parliament, set forth their grievances, and sue for remedy . . . yet, the more freely to carry out his rash designs, the King frequently commanded the sheriffs to cause certain persons nominated by himself to come to his Parliament; and the knights thus favorable to him he could, and frequently did, induce, sometimes by fear and divers threats, sometimes by gifts, to consent to measures prejudicial to the realm and excessively burdensome to the people; and especially he induced them to grant him the subsidy of wools for the term of his life, and another subsidy for a term of years, greatly oppressing the people.

Item, he unlawfully commanded that the sheriffs throughout the kingdom should swear beyond their ancient and accustomed oath to obey all his mandates under his signet whenever they should be addressed to them; and in case . . . they should hear of any persons . . . saying or repeating in public or in private anything tending to the discredit or slander of his person, to arrest them . . . and cause them to be imprisoned . . . until they should have further command from the King; as may be found by record. . . .

Item, in many Great Councils, when the lords of the realm, justices and others were charged faithfully to counsel the King in matters touching the estate of himself and the realm, the said lords, . . . when they gave counsel according to their discretion were often suddenly and so sharply rebuked and censured by him that they dared not . . . speak the truth in giving their advice.

Item, the King was wont almost continually to be so variable and dissembling in his words and writings, and so utterly contradictory, especially in writing to the Pope and to kings, and other lords within and without the realm, and to his other subjects, that scarcely any living man, being acquainted with his ways, could or would trust him. Indeed, he was held so faithless and inconstant that it gave ground for scandal not only as to his own person, but to the whole realm, and especially among foreigners throughout the world who became aware of it.

93. The Deposition of Richard II

The terms of Richard's abdication are here set forth: the complete renunciation of all royal power, the admission that he has deserved his fate, and the promise that he will abide by it. Actually, Richard was given little opportunity to break his word, for he died (he was probably murdered) the following year in Pontefract Castle, where he had been imprisoned.

In God's name, Amen. I, Richard, by the grace of God, King of England and France,[52] and Lord of Ireland, absolve all archbishops and bishops of the said kingdoms and lordships, and all other prelates whatsoever of secular or regular churches of whatsoever dignity, rank, state, or condi-

---

SOURCE: *Rotuli Parliamentorum* (London: Her Majesty's Stationery Office, 1776–77), *III*, 416–17, in George B. Adams and H. Morse Stephens, *Select Documents of English Constitutional History* (New York: Macmillan, 1937), 161–62.

---

[52] This claim to the French throne, asserted by Edward III, continued to appear in the royal style until the eighteenth century.

tion they may be, and dukes, marquises, earls, barons, knights, vassals, and vavassors[53] and all my liege men, clerical or secular by whatsoever name they are known, from the oath of fealty and homage and all others whatsoever made to me and from every bond of allegiance, royalty and lordship with which they have been or are bound by oath to me, or bound in any other way whatsoever; and these and their heirs and successors in perpetuity from these bonds and oaths and all other bonds whatsoever, I relieve, free, and excuse; absolved, excused and freed as far as pertains to my person, I release them from every performance of their oath which could follow from their promises or from any of them.

And all royal dignity and majesty and royalty and also the lordship and power in the said realms and lordship; and my other lordships and possessions or whatsoever others belong to me in any way, under whatsoever name they are known, which are in the aforesaid realms and lordships or elsewhere; and all right and color of right, and title, possessions, and lordship which I have ever had, still have or shall be able to have in any way, in these or any of them, or to these with their rights and everything pertaining to them or dependent upon them in any way whatsoever; from these or any of them; and also the command, government, and administration of such realms and lordships; and all and every kind of absolute and mixed sovereignty and jurisdiction in these realms and lordships belonging to me or to belong to me; the name and honor and royal right and title of king, freely, voluntarily, unequivocally, and absolutely, and in the best fashion, wise, and form possible, in these writings I renounce, and resign as a whole, and release in word and deed, and yield my place in them, and retire from them forever.

Saving to my successors, kings of England, in the realms and lordships and all other premises in perpetuity, the rights belonging or to belong to them, in them or in any of them, I confess, acknowledge, consider, and truly judge from sure knowledge that I in the rule and government of the said realms and lordships and all pertaining to them have been and am wholly insufficient and useless, and because of my notorious deserts am not unworthy to be deposed. And I swear on these holy Gospels touched bodily by me that I will never contravene these premises of renunciation, resignation, demise and surrender, nor will I impugn them in any way, in deed or in word by myself or by another or others, or as far as in me lies permit them to be contravened or impugned publicly or secretly, but I will hold this renunciation, resignation, demise, and surrender unalterable and acceptable and I will keep it firmly and observe it in whole and in every part; so may God help me and these holy scriptures of God. I, Richard, the aforesaid king, subscribe myself with my own hand.

[53] A class of feudal lords.

SECTION K

Parliament in the Early Fifteenth Century

94. The Regulation of Elections, 1406

Parliament, and particularly the House of Commons, refined its functions and procedures in the early fifteenth century – so much so that the period, coming before the governmental disorder of the middle years of the century, has sometimes been pictured as one of premature constitutional development. In 1406, because of improprieties in the election of knights of the shire, Parliament passed a regulatory statute.

. . . Our lord the King, at the grievous complaint of his Commons . . . of the undue election of the knights of counties for the Parliament, which be sometimes made through the prejudice of sheriffs, and otherwise against the form of the writs directed to the sheriff, . . . willing therein to provide remedy, by the assent of the Lords spiritual and temporal and of all the commonalty of the realm in this present Parliament, hath ordained and established that from henceforth the elections of such knights shall be made in the form as followeth: that is to say, that at the next county [court] to be held after the delivery of the writ . . . proclamation shall be made in the full county [court] of the day and place of the Parliament, and that all those that be there present . . . shall attend to the election of their knights for the Parliament; and then in the full county [court] they shall proceed to the election freely and indifferently,[54] notwithstanding any request or commandment to the contrary; and after they be chosen the names of the persons so chosen . . . shall be written in an indenture[55] under the seals of all them that did choose them, and fastened to the same writ of the Parliament; which indenture, so sealed and fastened, shall be held for the sheriff's return of the said writ touching the knights of the shires. . . .

SOURCES: *Statutes of the Realm* (London: Her Majesty's Stationery Office, 1810–28), *II*, 156.

Statutes of the Realm (London: Her Majesty's Stationery Office, 1810–28), *II*, 243.

---

[54] Impartially.

[55] A document, usually in duplicate, cut or notched so that the two parts would fit together exactly to prove authenticity.

95. A Property Qualification for Electors, 1429

Among the statutes governing parliamentary electoral procedure that of 1429 is particularly interesting in that it imposed a specific property qualification on electors. Knights of the shire were to be elected only by those persons possessed of freehold land bringing in an annual income of at least forty shillings – hence the expression, "forty-shilling freeholder." This act was not repealed until 1832.

. . . Whereas the elections of knights of the shire . . . in many counties of England have of late been made by a very large and excessive number of people dwelling within the same counties, of whom most were people of small substance or no value, but each of them pretending to have a voice equal, regarding such elections, to that of the most worthy knights and squires dwelling within the same counties; whereby manslaughters, riots, batteries and divisions among the gentlemen and other people . . . shall very likely occur, unless convenient remedy be provided in this behalf:

Our lord the King, considering the premises, hath provided and ordained by authority of this present Parliament that the knights of the shire . . . shall be chosen in every county by people dwelling and resident in the same, whereof every one of them shall have free land or tenement to the value of forty shillings by the year at the least above all charges; and that they who shall be so chosen shall be dwelling and resident within the same counties . . .

96. Requests of the Commons, 1401

In 1401 the House of Commons made two significant requests to the King. One to which the King agreed was that he not give ear to individual members of the House on matters being considered, but wait until the members as a whole could make their position known. In the other, which was not granted, they asked that the royal answers to their petitions be made known before parliamentary grants of money were made. They were clearly concerned with the position of their chamber as a deliberative assembly, and sought to insure the power of the purse in connection with governmental policies.

The . . . Commons set forth to our lord the King that, in connection with certain matters to be taken up among themselves, one of their num-

---

SOURCE: *Rotuli Parliamentorum* (London: Her Majesty's Stationery Office, 1776–77), *III*, 456–58, in Carl Stephenson and Frederick G. Marcham, *Sources of English Constitutional History* (New York: Harper and Row, 1937), 258–59.

ber, in order to please the King and to advance himself, might perchance tell our lord the King of such matters before they had been determined and discussed, or agreed on by the same Commons, whereby our same lord the King might be grievously moved against the said Commons or some of them. Therefore they very humbly prayed our lord the King that he would not receive any such person for the relating of any such matters. . . . To which it was responded on behalf of the King that he wished the same Commons to have deliberation and advisement in treating all their affairs among themselves, in order at their convenience to arrive at the best end and conclusion for the welfare and honor of him and of all his kingdom; and that he would not listen to any such person or give him credence until such matters had been presented to the King by the advice and consent of all the Commons, according to the purpose of their said petition. . . .

On the same Saturday the said Commons set forth to our said lord the King that in several Parliaments during times past their common petitions had not been answered before they had made their grant to our lord the King of some aid or subsidy. And therefore they prayed . . . that, for the great ease and comfort of the said Commons . . . [he] might be pleased to grant to the same Commons that they could have knowledge of the responses to their said petitions before any such grant had thus been made. . . .

97. Procedure Regarding Money Grants, 1407

In a dispute over a tax measure in 1407, which hinged on the role that the Commons should play, it was asserted by that chamber that it should make the grant, with the assent of the Lords. In other words, Commons proposed to initiate all money grants, and henceforth this became the practice. Bryce Lyon has called this "the key financial victory of the commons in the fifteenth century."[56]

It is to be remembered that on Monday, November 21, while our sovereign lord the King was in the council chamber within the Abbey of Gloucester, and while the lords spiritual and temporal assembled for this present Parliament were there before him, a conference was held among

SOURCE: *Rotuli Parliamentorum* (London: Her Majesty's Stationery Office, 1776–77), *III*, 611, in Carl Stephenson and Frederick G. Marcham, *Sources of English Constitutional History* (New York: Harper and Row, 1937), 263–65.

[56] Bryce Lyon, *A Constitutional and Legal History of Medieval England* (New York, Harper & Row, 1960), 603.

them regarding the state of the kingdom and the defence of the same. . . . And thereupon the aforesaid Lords were asked by way of interrogation what aid would be necessary and [how much] might suffice in this case. To which demand and question the same Lords severally responded that, in consideration on the one hand of the King's need and on the other of the poverty of his people, less aid would not suffice than a tenth and a half [a tenth] from the cities and boroughs and a fifteenth and a half [a fifteenth] from other laymen, and besides this an extension of the subsidy from wool, leather, and wool-fells, and of 3s the tun and 12d the pound,[57] from the feast of St. Michael next to the feast of St. Michael two years hence. Whereupon, by the command of our said lord the King, word was sent to the Commons of this present Parliament to cause a certain number of persons from their membership to appear before our said lord the King and the said Lords, in order to hear and to report to their companions what they might have by way of command from our said lord the King. And thereupon the said Commons sent into the presence of our said lord the King, and of the said Lords, twelve of their members, to whom, at the command of our same lord the King, the aforesaid question was set forth, together with the response severally made to it by the Lords aforesaid. Which response, it was the pleasure of our said lord the King, they should report to the rest of their companions in order that action might be taken to conform as nearly as possible with the opinion of the Lords aforesaid. When this report had thus been made to the said Commons, they were thereby greatly disturbed, saying and affirming that it was to the great prejudice and derogation of their liberties. And as soon as our said lord the King had heard this, desirous that neither at the present time nor in that to come anything should be done that in any way could be turned against the liberty of the estate, under which they had come to Parliament, or against the liberties of the Lords aforesaid, willed, granted, and declared, by the advice and consent of the same Lords, to the following effect: namely, that in the present Parliament and in all those to come the Lords may well confer among themselves in the King's absence regarding the state of the kingdom and the remedies thereby demanded, and that in the same way the Commons on their part may well confer among themselves with regard to the state and remedies aforesaid; always provided that neither the Lords . . . nor the Commons . . . shall make any report to our said lord the King of any grant made by the Commons and assented to by the Lords, or of the discussions in connection with the said grant, until the same Lords and Commons are of one mind and accord in this matter, and then . . . by the mouth of the Speaker of the said Commons

57 This was known as tunnage and poundage; the tun was a large cask.

. . . , so that the said Lords and Commons may hear the pleasure of our said lord the King. . . .

SECTION L

The Hundred Years' War: Fifteenth Century Climax

98. The Battle of Agincourt, 1415

Jehan de Waurin, a French nobleman who fought the English at Agincourt, gives this account of the most notable of Henry V's victories. The French losses numbered over 5000, the English, 200. Again, as under Edward III, the skill of the English bowmen was the chief factor in the overwhelming rout of the French.

Of the mortal battle of Agincourt, in which the King of England discomfited the French.

It is true that the French had arranged their battalions between two small thickets, one lying close to Agincourt, and the other to Tramecourt. The place was narrow, and very advantageous for the English, and, on the contrary, very ruinous for the French, for the said French had been all night on horseback, and it rained, and the pages, grooms, and others, in leading about the horses, had broken up the ground, which was so soft that the horses could with difficulty step out of the soil. And also the said French were so loaded with armor that they could not support themselves or move forward. In the first place they were armed with long coats of steel, reaching to the knees or lower, and very heavy, over the leg harness, and besides plate armor also most of them had hooked helmets; wherefore this weight of armor, with the softness of the wet ground, as has been said, kept them as if immovable, so that they could raise their clubs only with great difficulty, and with all these mischiefs there was this, that most of them were troubled with hunger and want of sleep. There was a marvellous number of banners, and it was ordered that some of them should be furled. Also it was settled among the said

---

SOURCE: Jehan de Waurin, *Chronicles, 1399–1422,* William Hardy (ed.), (London: Rolls Series, 1868), 210–213, in W. O. Hassall, *They Saw It Happen* (Oxford: Basil Blackwell, 1957), 192–94 (translated from the French by William Hardy and E. L. C. P. Hardy, Rolls Series, 1887).

French that everyone should shorten his lance, in order that they might be stiffer when it came to fighting at close quarters. They had archers and crossbowmen enough, but they would not let them shoot, for the plain was so narrow that there was no room except for the men-at-arms.

Now let us return to the English. After the parley between the two armies was finished, as we have said, and the delegates had returned, each to their own people, the King of England who had appointed a knight called Sir Thomas Erpingham to place his archers in front in two wings, trusted entirely to him, and Sir Thomas, to do his part, exhorted every one to do well in the same of the King, begging them to fight vigorously against the French in order to secure and save their own lives. And thus the knight, who rode with two others only in front of the battalion, seeing that the hour was come, for all things were well arranged, threw up a baton which he held in his hand, saying "Nestrocq,"[58] which was the signal for attack; then dismounted and joined the King, who was also on foot in the midst of his men, with his banner before him. Then the English, seeing this signal, began suddenly to march, uttering a very loud cry, which greatly surprised the French. And when the English saw that the French did not approach them, they marched dashingly towards them in very fine order, and again raised a loud cry as they stopped to take breath.

Then the English archers, who, as I have said, were in the wings, saw that they were near enough, and began to send their arrows on the French with great vigor. The said archers were for the most part in their doublets, without armor, their stockings rolled up to their knees, and having hatchets and battle-axes or great swords hanging at their girdles; some were bare-footed and bare-headed, others had caps of boiled leather, and others of osier, covered with harpoy or leather.

Then the French, seeing the English come towards them in this fashion, placed themselves in order, every one under his banner, their helmets on their heads. The constable, the marshal, the admirals, and the other princes earnestly exhorted their men to fight the English well and bravely; and when it came to the approach the trumpets and clarions resounded everywhere; but the French began to hold down their heads, especially those who had no bucklers, for the impetuosity of the English arrows, which fell so heavily that no one durst uncover or look up. Thus they went forward a little, then made a little retreat, but before they could come to close quarters, many of the French were disabled and wounded by the arrows; and when they came quite up to the English, they were, as has been said, so closely pressed one against another that none of them could lift their arms to strike against their enemies, except some that were

58 Perhaps "Now strike!"

in front, and these fiercely pricked with the lances which they had shortened to be more stiff, and to get nearer their enemies. . . .

99. The Treaty of Troyes, 1420

The Treaty of Troyes marked the zenith of English success in the Hundred Years' War. The following extracts from the treaty are to be read as the words of Charles VI, King of France, whose daughter, Catherine of Valois, Henry V married. Designed to effect an eventual union of England and France, it actually served to arouse a spirit of nationalism among the French.

First, that by the alliance of marriage made for the good of the said peace between our said son King Henry and our dearest and most beloved daughter Catherine, he has become our son and that of our very dear and best beloved companion, the Queen. . . .

Item, that after our death and afterwards the crown and the kingdom of France, together with its rights and appurtenances, shall be vested and remain perpetually in our son King Henry and his heirs. . . .

Item, that inasmuch as we are held and hindered for the present, on account of health, from hearing and arranging the affairs of the kingdom, the faculty and exercise of governing and ordering the public affairs of the said kingdom shall be vested, and remain during our life, in our said son King Henry, with the consent of the nobles and wise men obedient to us, who have at heart the advantage and honor of the said kingdom; and he shall be able to rule and govern by himself and by others whom he wishes to appoint, with the consent of the nobles and said wise men. . . .

Item, that during our life our said son King Henry will not call or subscribe himself, or cause himself to be named or written, as King of France, and will abstain utterly from this said name so long as we remain alive. . . .

Item, it is allowed that during our life we shall call, name, and write our said son, King Henry, in the French language in this manner: "Our dear son Henry, King of England, heir of France." . . .

Item, . . . the two crowns of France and England shall remain perpetually united, and shall be vested in one person, that is to say, in the person of our said son King Henry as long as he shall live, and after his death in the persons of his heirs who shall reign successively one after the other.

SOURCE: E. Cosneau, *Les Grands Traités de la Guerre de Cent Ans* (Paris: A Picard, 1889), 103–12, in Edward P. Cheyney, *Readings in English History from the Original Sources* (Boston: Ginn, 1935), 287–88.

Item, that now and forever all dissension, hatred, rancor, enmity, and war between the said kingdoms of France and England shall be allayed, suppressed, and stopped entirely, and that the people of these two kingdoms shall adhere to this said peace; and between the two said kingdoms there shall be agreement, now and forever, in the future to maintain peace, tranquillity, unity, and mutual affection, together with a strong and stable friendship; and that these two said kingdoms will help those who help them, and give mutual aid and assistance against all persons who bring violence, injury, grief, or loss to them, or to one of them; and that they talk matters over and contract freely and securely the one with the other in paying the customs and accustomed duties.

100. Joan of Arc and the Turning of the Tide

The Treaty of Troyes gave Henry V control of northern France, but the southern part of the country remained loyal to the son of Charles VI, despite his disinheritance. In 1422 both Henry V and Charles VI died. The former's brother, the Duke of Bedford, now directed the war in France with considerable success for several years. In 1429 Orleans, a key to the unconquered regions, was threatened. At this point Joan of Arc, a peasant girl, appeared on the scene. Convinced that she was sent by God, she evoked a patriotic fervor among the French which contributed to the eventual success of French arms. She actually led troops to the relief of Orleans, from which the English were compelled to withdraw. The following letter was dictated as she appeared before that city.

King of England, and you, Duke of Bedford, who call yourself Regent of the realm of France; you William de la Pole, Earl of Suffolk, John, Lord Talbot, and you Thomas, Lord Scales, who call yourselves lieutenants of the said Duke of Bedford: Do right to the King of heaven; deliver to the Maid, who is sent here by God, the King of heaven, the keys of all the good towns which you have taken and destroyed in France. She is come hither from God to restore the royal blood. She is ready to make peace if you are willing to do right to her, and on condition that you will quit France and pay back that which you have taken there. And you, archers, comrades of war, gentlemen and others who are before the city of Orleans, go away to your own country, in God's name; and if you do not thus, await tidings of the Maid, who will, ere long, come to see you to your very great hurt.

SOURCE: Jules Quicherat, *Procès de Condamnation et de Réhabilitation de Jeanne d'Arc* (Paris: J. Renouard, 1841), *I*, 240–41, in Edward P. Cheyney, *Readings in English History Drawn from the Original Sources* (Boston: Ginn, 1935), 292–93.

King of England, if you do not thus, I am the chief of the war, and in whatever place I shall find your men in France, I shall drive them out, whether they will or no; and if they will not obey, I will have them all slain.

I am sent here by God, the King of heaven, body for body, to drive you out of all France. And all who are willing to go I will receive to mercy. And do not have confidence in God, . . . for you shall not have the realm of France; but God, the King of heaven, wills that King Charles, the true heir, shall have it, and the Maid has revealed this to him. He shall enter Paris with a good company. If you will not believe the news that the Maid brings from God, in whatsoever place we find you we will attack you, and will make a greater slaughter than there has been in France in a thousand years, if you will not do right. And believe that the King of heaven will send more strength to the Maid than you can bring in all your assaults against her and her good soldiers. . . . You, Duke of Bedford, the Maid begs and requires you that you will not let yourself be destroyed. If you do right to her, you will be able to come into her company, wherein the French will do the best deed that ever was done for Christianity. And make answer, if you are willing to make peace in the city of Orleans; and if you do not do it, you will learn of it right soon to your very great loss. Written this Tuesday in Holy Week.

SECTION M

The Decline of Government under Henry VI

101. Jack Cade's Proclamation, 1450

In 1450 occurred Cade's Rebellion, so called because one of its leaders was an adventurer named Jack Cade. Caused, essentially, by dissatisfaction with the government of Henry VI, it differed from the Peasants' Revolt of 1381 in that "it had respectable upper middle-class support and was aimed less at landowners as such than at officials found to be oppressive."[59]

SOURCE: James Gairdner (ed.), *Three Fifteenth Century Chronicles* (London: Royal Historical Society, 1880), 94–98, in W. Garman Jones, *York and Lancaster* (London: G. Bell and Sons, 1914), 55–56.

---

[59] E. F. Jacob, *The Fifteenth Century, 1399–1485* (Oxford, 1961), 496–497.

These be the points, causes and mischiefs of gathering and assembling of us, the King's liege men of Kent, the 4th day of June the year of our Lord 1450, . . . the which we trust to Almighty God to remedy, with the help and the grace of God and of our sovereign lord the King and the poor commons of England, and else we shall die therefor:

We, considering that the King our sovereign lord, by the insatiable, covetous, malicious pomps, and false and of nought brought up certain persons,[60] that daily and nightly is about his Highness, and daily inform him that good is evil and evil is good. . . .

Item, they say that our sovereign is above his laws to his pleasure, and he may make it and break it as him list,[61] without any distinction. The contrary is true, and else he should not have sworn to keep it, the which we conceived for the highest point of treason that any subject may do to make his prince run into perjury.

Item, they say that the commons of England would first destroy the King's friends and afterwards himself, and then bring the Duke of York to be king. . . .

Item, they say the King should live upon his commons and that their bodies and goods be the King's; the contrary is true, for then needed him never Parliament to sit to ask good of his commons. . . .

Item, it is to be remedied that the false traitors will suffer no man to come into the King's presence for no cause without bribes where none ought to be had, nor no bribery about the King's person, but that any man might have his coming to him to ask him grace or judgment in such case as the King may give. . . .

Item, the law serveth of nought else in these days but for to do wrong. . . .

Item, we say our sovereign lord may understand that his false council hath lost his law, his merchandise is lost, his common people is destroyed, the sea is lost, France is lost, the King himself is so set that he may not pay for his meat and drink, and he oweth more than ever any King of England owed; for daily his traitors about him, where anything should come to him by his laws, anon they ask it from him. . . .

Item, his true commons desire that he will avoid from him all the false progeny and affinity of the Duke of Suffolk . . . and to take about his noble person his true blood of his royal realm, that is to say, the high and mighty prince the Duke of York, exiled from our sovereign lord's person by the noising of the false traitor, the Duke of Suffolk, and his affinity. . . .

Item, taking of wheat and other grains, beef, mutton and other victual, the which is unbearable hurt to the commons, without provision of our

60 Persons of obscure origins.

61 As he chooses.

sovereign lord and his true council, for his commons may no longer bear it.

Item, the statute upon the laborers and the great extortioners of Kent.[62] . . .

102. The Lack of Good Government, 1459

The evils of misgovernment are here sketched by the pen of a chronicler nearly a decade after the outbreak of Cade's Rebellion. In the meantime the Wars of the Roses, between the Lancastrian and Yorkist factions, had begun, contributing to an even more confused condition of affairs.

In this same time the realm of England was out of all good governance, as it had been many days before, for the King was simple and led by covetous counsel, and owed more than he was worth. His debts increased daily, but payment there was none; all the possessions and lordships that pertained to the crown the King had given away, some to lords and some to other simple persons, so that he had almost nought to live on. And such impositions as were put to the people, as taxes, tallages and quinzimes,[63] all that came from them were spent in vain, for he held no household nor maintained no wars. For these misgovernances, and for many other, the hearts of the people were turned away from them that had the land in governance, and their blessing was turned into cursing.

The Queen with such as were of her affinity ruled the realm as they liked, gathering riches innumerable. The officers of the realm, and especially the Earl of Wiltshire, Treasurer of England, for to enrich himself, peeled[64] the poor people and disinherited rightful heirs and did many wrongs. The Queen was defamed and slandered, that he that was called Prince was not her son. . . . Wherefore she, dreading that he should not succeed his father in the crown of England, allied unto her all the knights and squires of Cheshire, for to have their benevolence, and held open household among them . . . trusting them to make her son king. . . .

SOURCE: John S. Davies (ed.), *An English Chronicle of the Reigns of Richard II, Henry IV, Henry V and Henry VI* (London: Royal Historical Society, 1856), 79–80, in W. Garman Jones, *York and Lancaster* (London: G. Bell and Sons, 1914), 75.

---

[62] Several individuals are mentioned, including the Lord Treasurer, Crowmer, late Sheriff of Kent.

[63] Fifteenths; see p. 151.

[64] Plundered.

103. The Yorkist Claim Recognized, 1460

The factional strife characterizing the reign of Henry VI developed into a full-fledged civil war by the middle of the 1450's, by virtue of Richard of York's claim to the throne. Richard's military success was such that in 1460 he was able to secure formal recognition as Henry VI's heir, to the exclusion of Henry's seven-year-old son. Richard, however, was slain the same year at the Battle of Wakefield, and it remained for his son, Edward, to capitalize on his claim.

19. Blessed be Jesus, in whose hand and bounty resteth and is the peace and unity betwixt princes, and the weal of every realm, through whose direction agreed it is, appointed and accorded as followeth, betwixt the most high and most mighty prince King Henry the Sixth, King of England and of France, and Lord of Ireland, on the one part, and the right high and mighty prince Richard Plantagenet Duke of York, on the other part, upon certain matters of variance moved betwixt them, and in especial upon the claim and title unto the crowns of England and of France, and royal power, estate, and dignity, appertaining to the same, and lordship of Ireland, opened, showed and declared by the said Duke, before all the Lords spiritual and temporal being in this present Parliament: the said agreement, appointment and accord, to be authorized by the same Parliament. . . .

20. The said title nevertheless notwithstanding, and without prejudice of the same, the said Richard Duke of York, tenderly desiring the weal, rest and prosperity of this land, and to set apart all that might be trouble to the same; and considering the possession of the said King Henry the Sixth, and that he hath for his time been named, taken and reputed King of England and of France, and Lord of Ireland; is content, agreed and consenteth, that he be had, reputed and taken, King of England and of France, with the royal estate, dignity and pre-eminence belonging thereto, and Lord of Ireland, during his life natural; and for that time, the said Duke without hurt or prejudice of his said right and title, shall take, worship and honor him for his sovereign lord.

21. *Item,* the said Richard Duke of York, shall promise and bind him by his solemn oath, in manner and form as followeth:

In the name of God, Amen. I Richard Duke of York, promise and swear by the faith and truth that I owe to almighty God, that I shall never do, consent, procure or stir, directly or indirectly, privily or openly, nor as much as in me is and shall be, suffer to be done, consented, pro-

---

SOURCE: *Rotuli Parliamentorum* (London: Her Majesty's Stationery Office, 1776–77), V, 378–79.

cured or stirred, anything that may be . . . to the abridgement of the natural life of King Henry the Sixth, or to the hurt or diminishing of his reign or dignity royal, by violence or any other wise, against his freedom and liberty; but that if any person or persons would do or presume anything to the contrary, I shall with all my power and might withstand it, and make it be withstood as far as my power will stretch thereunto; so help me God, and these holy evangelists.

Item, Edward Earl of March, and Edmund Earl of Rutland, the sons of the said Richard Duke of York, shall make like oath.

22. *Item,* it is accorded, appointed and agreed, that the said Richard Duke of York enjoy, be entitled, called and reputed henceforth very and rightful heir to the crowns, royal estate, dignity and lordship abovesaid; and after the decease of the said King Henry, or when he will lay from him the said crowns, estate, dignity and lordship, the said Duke and his heirs shall immediately succeed to the said crowns, royal estate, dignity and lordship.

SECTION N

The Yorkists in Power

104. Fortescue's Prescription for Strong Monarchical Government

Edward, Duke of York, secured the throne following his victory over the Lancastrians at Towton in 1461. Save for one brief interlude (1470–1471) he retained it until his death in 1483. While the country was not entirely free of the disorders which had plagued it under Henry VI, the royal authority was much more effective. This was in large measure due to the improved financial condition of the crown, to which Edward's own policies markedly contributed. The contemporary judge and political writer, Sir John Fortescue, noted in his *Governance of England* the disadvantages arising from "over-mighty" subjects and a financially embarrassed monarch.

Then needs it that the king's livelihood, above such revenues as shall be assigned for his ordinary charges, be greater than the livelihood of

---

SOURCE: William H. Dunham and Stanley Pargellis, *Complaint and Reform in England, 1436–1714* (New York: Oxford University Press, 1938), 64-66.

the greatest lord in England. And peradventure, when livelihood sufficient for the king's ordinary charges is fixed and assigned thereto, it shall appear that divers lords of England have also much livelihood of their own, as then shall remain in the king's hands for his extraordinary charges; which were inconvenient and would be to the king right dreadful. For then such a lord may spend more than the king, considering that he is charged with no such charges, extraordinary or ordinary as is the king, except an household, which is but little in comparison of the king's house. Wherefore if it be thus, it shall be necessary that there be purveyed for the king much greater livelihood than he has yet. For man's courage is so noble, that naturally he aspires to high things, and to be exalted, and therefore enforces himself to be always greater and greater. For which the philosopher says, *omnia amamus sed principari maius.*[65] Whereof it has come that oftentimes when a subject has had also great livelihood as his prince, he has anon aspired to the estate of his prince, which by such a man may soon be got. For the remnant of the subjects of such a prince, seeing that if so mighty a subject might obtain the estate of their prince, they should then be under a prince double so mighty as was their old prince – which increase any subject desires, for his own discharge of that he bears to the sustenance of his prince – and therefore will be right glad to help such a subject in his rebellion. And also such an enterprise is the more feasible, when such a rebel has more riches than his sovereign lord. For the people will go with him that best may sustain and reward them. This manner of doing has been so often practised almost in every realm that their chronicles be full of it. . . . We have also seen late in our realm some of the king's subjects give him battle, by occasion that their livelihood and offices were the greatest of the land, and else they would not have done so. . . .

And also it may not be eschewed,[66] but that the great lords of the land by reason of new descents[67] falling unto them, by reason also of marriages, purchases, and other titles, shall oftentimes grow to be greater than they be now, and peradventure some of them to be of livelihood and power like a king; which shall be right good for the land while they aspire to none higher estate. For such was the Duke of Lancaster, that warred the King of Spain, one of the mightiest kings of Christendom, in his own realm. But this is written only to the intent that it be well understood how necessary it is that the king have great possessions and peculiar livelihood for his own surety; namely, when any of his lords shall happen to be so excessively great as there might thereby grow peril

[65] We love all things, but most of all to rule.
[66] Avoided.
[67] Inheritances.

to his estate. For certainly there be no greater peril grow to a prince than to have a subject equally powerful to himself.

105. Edward IV's Financial Arrangements

Edward IV was more independent of Parliament than the Lancastrian kings had been, only six Parliaments assembling in his reign. As already noted, this was largely owing to the success of his financial policies, a subject dealt with by the contemporary continuer of the Croyland Chronicle.

There is no doubt that the King . . . was by no means ignorant of the condition of his people, and how readily they might be betrayed, in case they should find a leader, to enter into rebellious plans and conceive a thirst for change. Accordingly, seeing that things had now come to such a pass that from henceforth he could not dare, in his emergencies, to ask the assistance of the English people, . . . he turned all his thoughts to the question: how he might in future collect an amount of treasure worthy of his royal station out of his own substance and by the exercise of his own energies.

Accordingly, having called Parliament together, he resumed possession of nearly all royal estates, without regard to whom they had been granted, and applied the whole thereof to the support of the expenses of the crown. Throughout all the ports of the kingdom he appointed inspectors of the customs, men of remarkable shrewdness, but too hard, according to general report, upon the merchants. The King himself also, having procured merchant ships, put on board of them the finest wools, cloths, tin and other productions of the kingdom, and like a private individual living by trade exchanged merchandise for merchandise by means of his factors, among both Italians and Greeks. The revenues of vacant prelacies, which according to Magna Carta cannot be sold, he would only part with out of his hands at a stated sum, and on no other terms whatever. He also examined the register and rolls of Chancery, and exacted heavy fines from those whom he found to have intruded and taken possession of estates without prosecuting[68] their rights in form required by law, by way of return for the rents which they had in the meantime received.

These, and more of a similar nature than can possibly be conceived by

---

SOURCE: Henry T. Riley (ed.), *Ingulph's Chronicle of the Abbey of Croyland, with the Continuations* (London: G. Bell and Sons, 1854), 474–75.

---

[68] Seeking to enforce.

a man who is inexperienced in such matters, were his methods of making up a purse; added to which there was the yearly tribute of £10,000 due from France, together with numerous tenths from the churches, from which the prelates and clergy had been unable to get themselves excused. All these particulars, in the course of a very few years, rendered him an extremely wealthy prince. . . .

106. Richard of Gloucester Seizes the Crown, 1483

Upon Edward IV's death in 1483 the throne passed to his son Edward, who was but thirteen years of age. He reigned less than three months; his uncle, Richard, Duke of Gloucester, first established himself as Protector, then seized the crown itself, ruling for two years as Richard III. Dominic Mancini, an Italian cleric and humanist visiting England at the time, provides this comment on Richard's usurpation.

When Richard felt secure from all those dangers that at first he feared, he took off the mourning clothes that he had always worn since his brother's death, and putting on purple raiment he often rode through the capital surrounded by a thousand attendants. He publicly showed himself so as to receive the attention and applause of the people as yet under the name of Protector; but each day he entertained to dinner at his private dwellings an increasingly large number of men. When he exhibited himself through the streets of the city he was scarcely watched by anybody, rather did they curse him with a fate worthy of his crimes, since no one now doubted at what he was aiming. . . .

In the meantime the Duke summoned to London all the peers of the realm: the latter supposed they were called both to hear the reason of Hastings's[69] execution, and to decide again about the coronation of Edward. . . . Each came with the retinue that his title and station demanded; but the Duke advised them to retain a few attendants, who were indispensable for their personal service, and to send back the others to their own homes. . . . They obeyed his instructions, and when the Duke saw that all was ready, as though he knew nothing of the affair, he secretly dispatched the Duke of Buckingham to the lords with orders to submit to their decision the disposal of the throne. He argued that it appeared unjust that this lad, who was illegitimate, should assume the office of kingship; for he was a bastard by reason of his father Edward

SOURCE: C. A. J. Armstrong (ed.), *The Usurpation of Richard III* (Oxford: The Clarendon Press, 1936), 115–21.

[69] Baron Hastings, beheaded on a charge of treason in 1483.

having married Elizabeth,[70] when by law he was contracted to another wife, whom the Duke of Warwick had given him. . . . As for the son of the Duke of Clarence, he had been rendered ineligible for the crown by the felony of his father, since his father after conviction for treason had forfeited not only his own but also his son's right of succession. The only survivor of the royal stock was Richard, Duke of Gloucester, who was legally entitled to the crown, and could bear its responsibilities thanks to his proficiency. His previous career and blameless morals would be a sure guarantee of his good government. Although he would refuse such a burden, he might yet change his mind if he were asked by the peers.

On hearing this the lords consulted their own safety, warned by the example of Hastings, and perceiving the alliance of the two dukes, whose power, supported by a multitude of troops, would be difficult and hazardous to resist. . . . They saw themselves surrounded and in the hands of the dukes, and therefore they determined to declare Richard their king and ask him to undertake the burden of office. On the following day all the lords forgathered at the house of Richard's mother, whither he had purposely betaken himself, that these events might not take place in the Tower, where the young King was confined.

There the whole business was transacted, the oaths of allegiance given, and other indispensable acts fittingly performed. On the two following days the people of London and the heads of the clergy did likewise. . . . This being accomplished, a date was fixed for the coronation, while acts in the name of Edward V since the death of his father were repealed or suspended, seals and titles changed, and everything confirmed and carried on in the name of Richard III.

SECTION O

The Regulation of Trade and Industry

107. The Staplers' Charter, 1313

The fourteenth and fifteenth centuries witnessed the development of national, as opposed to local (guild), policies affecting commerce and

SOURCE: A. E. Bland, P. A. Brown and R. H. Tawney, *English Economic History: Select Documents* (London: G. Bell and Sons, 1930), 178–80.

[70] Elizabeth Woodville, privately married to Edward IV in 1464.

industry. Many of these policies were mercantilist in nature, that is, they were designed to develop the economic self-sufficiency of the state. One device was the creation of staples (places through which trade in certain commodities was to be channeled), which came into being in the thirteenth century. Since production and export of wool was, next to agriculture, the leading industry, the government took a particular interest in controlling this activity. The following document, dated 1313, came to be known as the Staplers' Charter.

The King to all to whom, etc., greeting. Know ye that whereas before these times divers damages and grievances in many ways have befallen the merchants of our realm, not without damage to our progenitors, sometime Kings of England, and to us, because merchants, as well denizen as alien, buying wools and wool-fells within the realm aforesaid and our power, have gone at their pleasure with the same wools and fells, to sell them, to divers places within the lands of Brabant, Flanders and Artois:

We, wishing to prevent such damages and grievances and to provide as well as we may for the advantage of us and our merchants of the realm aforesaid, do will and by our Council ordain, to endure for ever, that merchants denizen and alien, buying such wools and fells within the realm and power aforesaid and wishing to take the same to the aforesaid lands to sell there, shall take those wools and fells or cause them to be taken to a fixed staple to be ordained and assigned within any of the same lands by the mayor and community of the said merchants of our realm, and to be changed as and when they shall deem expedient, and not to other places in these lands in any wise.

Granting to the said mayor and merchants of our realm aforesaid, for us and our heirs, that the mayor and council of the same merchants for the time being may impose upon all merchants, denizen and alien, who shall contravene the said ordinance and shall be reasonably convicted thereof by the aforesaid mayor and council of the said merchants, certain money penalties for those offences, and that such money penalties, whereof we or our ministers shall be informed by the aforesaid mayor, shall be levied to our use from the goods and wares of merchants so offending, wheresoever they shall be found within the realm and power aforesaid, by our ministers, according to the information aforesaid and the assessment thereof to be made by the mayor himself, saving always to the said mayor and merchants that of themselves they may reasonably chastise and punish offending merchants, if their goods and wares chance to be found in the staple aforesaid outside our realm and power aforesaid, without interference or hindrance on the part of us or our heirs or our ministers whomsoever, as they have hitherto been wont to do. In witness whereof, etc. Witness the King at Canterbury, 20 May.

By the King himself.

108. Prohibition of Export of Raw Materials, 1326

Restrictions on the export of raw materials were a prominent feature of English policy, on the grounds that the national economy benefitted more when such commodities were worked up by native artisans. In 1326 the following order prohibiting the export of materials for making cloth was issued by Edward II:

Edward by the grace of God, King of England, etc., to our well-beloved Hamon de Chigewelle, Mayor of our city of London, greeting. We have read the letters that you have sent us, in the which you have signified unto us that Flemings, Brabanters and other aliens have been suddenly buying throughout our land all the teasels[71] that they can find; and are also buying butter, madder, woad, fuller's earth, and all other things which pertain to the working of cloth, in order that they may disturb the staple and the common profit of our realm; and further, that you have stopped twenty tuns that were shipped and ready for going beyond sea, at the suit of good folks of our said city; upon your doing the which we do congratulate you, and do command and charge you, that you cause the said tuns well and safely to be kept; and if any such things come into our said city from henceforth, to be sent beyond the sea by merchants, aliens or denizens, cause them also to be stopped and safely kept, until you shall have had other mandate from us thereon; and you are not to allow any such things to pass through your bailiwick, by reason whereof the profit of our staple may be disturbed. We have also commanded our Chancellor, that by writs under our Great Seal he shall cause it everywhere to be forbidden that any such things shall pass from henceforth out of our realm, in any way whatsoever. Given under our Privy Seal at Saltwood, the 21st day of May, in the 19th year of our reign.

109. Export of Gold and Silver Prohibited, 1423

Emphasis on the retention of precious metals within the realm, a prominent characteristic of mercantilist thought, is seen in this statute of Henry VI.

SOURCES: Henry T. Riley, *Memorials of London and London Life in the Thirteenth, Fourteenth and Fifteenth Centuries* (London: Longmans, Green, 1868), 149–50.

Statutes of the Realm (London: Her Majesty's Stationery Office, 1810–28), *II*, 219–20.

---

[71] A prickly herb used by fullers to raise a nap on cloth.

It is ordained and established that no gold nor silver shall be carried out of the realm . . . unless it be for payment of wars and the King's soldiers beyond the sea, . . . except the ransoms for fines of English prisoners taken and to be taken beyond the sea, and the money that the soldiers shall carry with them for their reasonable costs, and also for horses, oxen, sheep and other things bought in Scotland, to be sent and carried to the parties adjoining. . . .

And because it is supposed that the money and gold of the realm is carried out of the same by merchants aliens, it is ordained and established that the merchants aliens shall find surety in the Chancery, every company for them of their company, that none of them shall carry out of the realm any gold or silver against the form of the said statute, upon pain of forfeiture of the same gold or silver, or the value of the same. . . .

110. Imports Restricted, 1463

In order to increase opportunities of employment for English artisans, and hence alleviate distress and augment national prosperity, acts were passed prohibiting the importation of manufactured articles. Here is the preamble of such an act; among the prohibited goods were woolen cloth, laces, ribbons, embroidered silk, saddles, harness, locks, purses, gloves, hats, knives and brushes.

Whereas in . . . Parliament, by the artificers of manual occupations, men and women, inhabiting and resident in the city of London and other cities, towns, boroughs and villages within this realm of England and Wales, it hath been piteously showed and complained how that they all in general, and every one of them, be greatly impoverished, and much hindered and prejudiced of their worldly increase and daily living, by the great multitude of divers commodities and wares pertaining to their mysteries[72] and occupations, being fully wrought and ready made for sale, as well by the hands of strangers being the King's enemies as other in this realm and Wales, fetched and brought from beyond the sea . . . whereof the greatest part in substance is deceitful, and nothing worth in regard of any man's occupation or profit; by which occasion the said artificers cannot live by their mysteries and occupations, as they have done in times past, but divers of them . . . be at this day unoccupied, and do

SOURCE: *Statutes of the Realm* (London: Her Majesty's Stationery Office, 1810–28), *II*, 396.

[72] Crafts, trades.

hardly live in great misery, poverty and need, whereby many inconveniences have grown before this time, and hereafter more be like to come . . . if due remedy be not in this behalf provided. . . .

111. The Regulation of Cloth-making, 1465

The national government took steps to impose standards of industrial craftsmanship and to regulate conditions of employment, as is seen in this statute of 1465.

Our lord King Edward the Fourth . . . by the advice and assent of the Lords spiritual and temporal, and at the special request of his Commons, . . . hath ordained and established certain statutes and ordinances in form following.

First, whereas many years past, and now at this day, the workmanship of cloth, and things requisite to the same, is and hath been of such fraud, deceit and falsity that the said cloths in other lands and countries be had in small reputation, to the great shame of this land, and by reason thereof a great quantity of cloths of other strange lands be brought into this realm, and there sold at an high and excessive price . . . [regulations are imposed to insure a standard of production].

Also whereas before this time in the occupations of cloth-making, the laborers thereof have been driven to take a great part of their wages in pins, girdles and other unprofitable wares at a price less than their lawful wages, and also have delivered to them wools to be wrought by very excessive weight, whereby both men and women have been discouraged of such labor:

Therefore it is ordained . . . that every man and woman being cloth-makers . . . shall pay to the carders, spinsters and all such other laborers in any part of the said trade, lawful money for all their lawful wages . . . and also shall deliver wools to be wrought according to the faithful delivery and due weight thereof, upon pain of forfeiture to the same laborer of the treble of his said wages so unpaid, as often as the said cloth-maker doth refuse to pay . . .

Also it is ordained . . . that every carder, spinster, weaver, fuller, shearman and dyer shall duly perform his duty in his occupation, upon pain to yield to the party grieved in this behalf his double damages. . . .

Also it is ordained . . . that all manner of woolen cloths made in any other region, brought into this realm of England and set to sale within

---

SOURCE: *Statutes of the Realm* (London: Her Majesty's Stationery Office, 1810–28), *II*, 403–07.

any part of this realm . . . shall be forfeit to our said sovereign lord the King, except cloths made in Wales and Ireland, and cloths taken by any of the King's liege people upon the sea, without fraud or collusion. . . .

112. Apprenticeship

Though craft guilds declined in the latter part of the middle ages, apprenticeship remained a prominent feature of medieval industry. It was designed to provide the necessary training in some particular craft, or industrial pursuit. To some extent the master of the apprentice stood in the position of a parent and was responsible for his behavior and upbringing. The relationship between apprentice and master was set forth in a document called an indenture, of which the following (dated 1459) may serve as an example.

This indenture made between John Gibbs of Penzance in the County of Cornwall of the one part and John Goffe, Spaniard, of the other part, witnesses that the aforesaid John Goffe has put himself to the aforesaid John Gibbs to learn the craft of fishing and to stay with him as apprentice and to serve from the feast of Philip and James next to come after the date of these presents until the end of eight years then next ensuing and fully complete; throughout which term the aforesaid John Goffe shall well and faithfully serve the aforesaid John Gibbs and Agnes his wife as his masters and lords, shall keep their secrets, shall everywhere willingly do their lawful and honorable commands, shall do his masters no injury, nor see injury done to them by others, but prevent the same, as far as he can, shall not waste his master's goods nor lend them to any man without his special command. And the aforesaid John Gibbs and Agnes his wife shall teach, train and inform or cause the aforesaid John Goffe, their apprentice, to be informed in the craft of fishing in the best way they know, chastising him duly, and finding for the same John, their apprentice, food, clothing, linen and woollen, and shoes, sufficiently, as befits such an apprentice to be found, during the term aforesaid. And at the end of the term aforesaid the aforesaid John Goffe shall have of the aforesaid John Gibbs and Agnes his wife twenty shillings sterling without any fraud. In witness whereof the parties aforesaid have interchangeably set their seals to the parts of this indenture. These witnesses: Richard Bascawen, Robert Martyn and Robert Cosyn and many others. Given at Penzance, 1 April in the thirty-seventh year of the reign of King Henry the Sixth after the conquest of England.

SOURCE: A. E. Bland, P. A. Brown and R. H. Tawney, *English Economic History: Select Documents* (London: G. Bell and Sons, 1930), 147–48.

113. A Petition against Usury

Not until near the end of the fifteenth century were legal rates of interest fixed by the state. The practice of lending money at interest, or usury, was condemned by the medieval church, and as late as the early seventeenth century English ecclesiastical courts still punished the offense. But even before the expulsion of the Jews, once the acknowledged money-lenders of the realm, Christians were getting around the prohibition in various ways. The following petition reveals the opposition of the Commons to usury in 1376.

Further, the Commons of the land pray that whereas the horrible vice of usury is so spread abroad and used throughout the land that the virtue of charity, without which none can be saved, is wellnigh wholly perished, whereby, as is known too well, a great number of good men have been undone and brought to great poverty:

Please it, to the honor of God, to establish in this present Parliament that the ordinance[73] made in the city of London for a remedy of the same, well considered and corrected by your wise Council and likewise by the Bishop of the same city, be speedily put into execution, without doing favor to any, against every person, of whatsoever condition he be, who shall be hereafter attainted as principal or receiver or broker of such false bargains. And that all the mayors and bailiffs of cities and boroughs throughout the realm have the same power to punish all those who shall be attainted of this falsity within their bailiwicks according to the form of the articles comprehended in the same ordinance. And that the same ordinance be kept throughout all the realm, within franchises[74] and without.

SOURCE: A. E. Bland, P. A. Brown and R. H. Tawney, *English Economic History: Select Documents* (London: G. Bell and Sons, 1930), 200–01.

---

[73] An ordinance dated 1363.
[74] Privileged jurisdictions, characteristic of feudal society.

Part 4

❖❖

THE TUDOR ERA

1485-1603

SECTION A

Henry VII and the Ordered Realm

✣✣

114. Lawlessness in 1485

The lawlessness which prevailed throughout so much of the fifteenth century was still a problem when Henry VII, first of the Tudor sovereigns, came to the throne in 1485. Reference to this condition is to be found in a contemporary Year Book, or law report.

And after dinner, all the Justices were at Blackfriars[1] to discuss the King's business for the Parliament. And several good statutes were mentioned, very advantageous for the kingdom if they could be carried out. These were the statutes compiled in the time of Edward IV and sent into each county to the justices of the peace, to be proclaimed and enforced, viz., Winchester and Westminster[2] for robberies and felonies, the statute of riots, routs and forcible entry, the statute of laborers and vagabonds, of tokens, and liveries, maintenance and embracery. And now they agreed that the Statute 23 Henry VI concerning sheriffs, etc., should be sent to them and then they would have enough, and if they were properly carried out the law would run its course well. But the question was, would they be carried out. And the Chief Justice said that the law would never be properly carried out until the lords spiritual and temporal are of one mind, for the love and fear they have of God or the

---

SOURCE: C. H. Williams, *England under the Early Tudors, 1485–1529* (London: Longmans, Green, 1925), 169–70.

---

[1] A Dominican monastery in London.
[2] Statutes of the reign of Edward I.

King or both, to carry them out effectively. Thus, when the King on his side and the lords on theirs will do this, every one else will quickly do it, and if they do not they will be punished, and then all will be warned (by their example). For he said that in the time of Edward IV, when he was Attorney, he saw all the lords sworn to keep and execute diligently the statutes which they with others had just drawn up by command of the King himself. And within an hour, while they were still in the Star Chamber, he saw the lords making retainers by oath, and swearing, and doing other things contrary to their above-mentioned promises and oaths. Consequently oaths and swearing are of no use until they are in the aforesaid mind. And he said that he had told this to the King himself.

115. The Court of Star Chamber

To obtain obedience to the law from powerful men, who frequently ignored or interfered with the regular courts of justice, Henry VII in 1487 by statute allotted special functions to the Council. Those exercising this jurisdiction came in time to be regarded as constituting a separate agency, the Court of Star Chamber, so called because of the decor of its ceiling.

The King, our sovereign lord, remembereth how, by unlawful maintenances, giving of liveries, signs and tokens, and retainders by indenture, promises, oaths, writing or otherwise,[3] embraceries[4] of his subjects, untrue demeanings of sheriffs in making of panels and other untrue returns, by taking of money by juries, by great riots and unlawful assemblies, the policy and good rule of this realm is almost subdued, and for the none punishment of this inconvenience and by occasion of the premises nothing or little may be found by inquiry; whereby the laws of the land in execution may take little effect, to the increase of murders, robberies, perjuries, and unsureties of all men living, and losses of their lands and goods, to the great displeasure of almighty God:

Be it therefore enacted for reformation of the premises by the authority of this Parliament that the Chancellor and Treasurer of England for the time being, and Keeper of the King's Privy Seal, or two of them, calling to [them] a bishop and temporal lord of the King's most honor-

---

SOURCE: *Statutes of the Realm* (London: Her Majesty's Stationery Office, 1810–28), *II*, 509–10.

---

[3] The reference here is to various means of mainaining retainers; on liveries, see p. 175.
[4] Embracery: influencing, or attempting to influence, a court or jury illegally.

able Council and the two chief justices of the King's Bench and Common Pleas for the time being, or other two justices in their absence, upon bill or information put to the said Chancellor, for the King or any other, against any person for any misbehaving afore-rehearsed, have authority to call before them by writ or privy seal the said misdoers, and them and other by their discretions to whom the truth may be known to examine, and such as they find therein defective to punish them after their demerits, after the form and effect of statutes thereof made. . . .

116. The Statute of Liveries, 1504

Henry VII fully realized the importance of reducing the power of the "overmighty" subjects. One method was to take steps against the giving of liveries, whereby bodies of retainers were created. At his first Parliament Henry forced the lords to take an oath against the practice. But it died hard, for Parliament saw fit to legislate against it nearly twenty years later. The term "livery" was first applied to allowances given to servants; later it came to connote a distinctive garb, or heraldic badge, worn by them.

The King our sovereign lord calleth to his remembrance that, where before this time divers statutes for punishment of such persons that give or receive liveries, or that retain any person or persons, . . . have been made and established, and that notwithstanding divers persons have taken upon them, some to give and some to receive liveries and to retain and be retained, contrary to the form of the said statutes, and little or nothing is or hath been done for the punishment of the offenders in that behalf:

Wherefore our sovereign lord the King, by the advice of the Lords spiritual and temporal and of his Commons of his realm in this Parliament . . . hath ordained, established and enacted that all his statutes and ordinances afore this time made . . . be . . . put in due execution. And . . . the King ordaineth . . . by the said authority that no person, of what estate or degree or condition he be . . . , privily or openly give any livery or sign or retain any person, other than such as he giveth household wages unto without fraud or color, or that he be his manual servant or his officer or man learned in the one law or the other, by any writing, oath, promise, livery, sign, badge, token or in any other manner . . . unlawfully retain; and if any do the contrary, that then he run and fall in the pain and forfeiture for every such livery and sign, badge or token,

---

SOURCE: *Statutes of the Realm* (London: Her Majesty's Stationery Office, 1810–28), *II*, 658.

100s, and the taker and accepter of every such livery, badge, token or sign to forfeit and pay for every livery and sign, badge or token so accepted, 100s. . . .

SECTION B

Signs of the Renaissance

117. England as Seen by Erasmus

By the late fifteenth century there is increasing evidence that England was being affected by the intellectual and artistic currents of the Italian Renaissance. These two letters of the great Dutch humanist, Erasmus, both written from England, give his impressions of the state of the "new" learning there, the first being dated 1499, the second 1518. Note his reference to John Colet, William Grocyn, Thomas Linacre and Thomas More, the leading English humanists of the time.

Erasmus to Robert Fisher[5] **(1499)** – I have been rather afraid of writing to you, dearest Robert, not that I feared your affection had been at all lessened by such distances of time and place, but because you are in a country where the walls are more learned and eloquent than our men; so that what we here think eloquent and beautiful cannot but seem poor and rude and tasteless there. Your England naturally expects you to return not only most learned in the laws but equally loquacious in Greek and Latin. You would have seen me, too, in Italy before this time if my lord Mountjoy, when I was prepared for the journey, had not carried me off to England. . . .

But how do you like our England, you will say. Believe me, my Robert, when I answer that I never liked anything so much before. I find the climate both pleasant and wholesome; and I have met with so much kindness and so much learning – not hackneyed and trivial, but deep, accurate, ancient Latin and Greek – that but for the curiosity of seeing it, I do not now care so much for Italy. When I hear my Colet, I seem to

---

SOURCE: F. M Nichols, *The Epistles of Erasmus* (London: Longmans, Green, 1901–18), *I*, 225–26; *III*, 345.

---

[5] Robert Fisher, English agent in Italy. A pupil of Erasmus, he had some experience in diplomatic affairs and held some church preferments.

be listening to Plato himself. In Grocyn, who does not marvel at such a perfect round of learning? What can be more acute, profound, and delicate than the judgment of Linacre? What has nature ever created more gentle, more sweet, more happy than the genius of Thomas More? I need not go through the list. It is marvelous how general and abundant is the harvest of ancient learning in this country, to which you ought all the sooner to return. . . .

Erasmus to Richard Pace[6] **(1518)** – Your King's court in Britain is brilliant indeed, the seat and citadel of the best studies and of the highest characters! I congratulate you . . . upon having such a sovereign, and I congratulate the Prince himself, whose reign is made illustrious by so many lights of genius; and on both accounts I congratulate your England, a fortunate country in many ways besides, but so excelling in these respects that no region in the world can be compared with it. Now at any rate a whole lifetime may be spent with advantage in a country where under princely favor good letters are dominant, the love of honor is strong, and a sentence of banishment has been passed against that futile and tasteless learning with its masked affectation of holiness, which used to be in fashion with uneducated men of education. I grieve to hear that Grocyn is failing; while I see that in place of one learned scholar so many will soon grow up. . . .

118. The Education of Gentlemen

With the advent of the sixteenth century "the ideal of education changed from the theological to the rhetorical, from the training of priests and scholars to the training of accomplished gentlemen serving the state."[7] Particularly influential in this regard was Sir Thomas Elyot's *The Governour* (1531), the first book on education to be published in English. The following excerpt (as well as other passages in the work) shows the high regard for Greek and Roman classics, which are given priority even over the Scriptures.

By the time that the child do come to seventeen years of age, to the intent his courage be bridled with reason, it were needful to read unto him some works of philosophy, specially that part that may inform him

---

SOURCE: Thomas Elyot, *The Governour* (New York: Everyman's Library, J. M. Dent and Sons, 1937), 47–48.

---

[6] Diplomatist and Dean of St. Paul's Cathedral, London.
[7] G. R. Elton, *England under the Tudors* (London, 1960), 431.

unto virtuous manners, which part of philosophy is called moral. Wherefore there would be read to him, for an introduction, two [of] the first books of the work of Aristotle, called *Ethicae,* wherein is contained the definitions and proper significations of every virtue; and that to be learned in Greek, for the translations that we yet have be but a rude and gross shadow of the eloquence and wisdom of Aristotle. Forthwith would follow the work of Cicero, called in Latin *De Officiis,* whereunto yet is no proper English word to be given; but to provide for it some manner of exposition it may be said in this form: "Of the Duties and Manners Appertaining to Men." But above all other, the works of Plato would be most studiously read when the judgment of a man is come to perfection, and by the other studies is instructed in the form of speaking that philosophers used.

Lord God, what incomparable sweetness of words and matter shall he find in the said works of Plato and Cicero, wherein is joined gravity with delectation, excellent wisdom with divine eloquence, absolute virtue with pleasure incredible, and every place is so enforced with profitable counsel, joined with honesty, that those three books be almost sufficient to make a perfect and excellent governor.

The Proverbs of Solomon, with the books of Ecclesiastes and Ecclesiasticus,[8] be very good lessons. All the historical parts of the Bible be right necessary for to be read of a nobleman, after that he is mature in years. And the residue (with the New Testament) is to be reverently touched, as a celestial jewel or relic. . . .

It would not be forgotten that the little book of the most excellent Doctor Erasmus Roterodamus[9] (which he wrote to Charles, now being Emperor and then Prince of Castile), which book is entitled *The Institution of a Christian Prince,* would be as familiar always with gentlemen, at all times and in every age, as was Homer with the great king Alexander, or Xenophon with Scipio; for, as all men may judge that have read that work of Erasmus, there was never a book written in Latin that, in so little a portion, contained of sentence, eloquence and virtuous exhortation a more compendious abundance. . . .

[8] One of the Apocrypha, or Old Testament books, rejected by Protestants.
[9] Erasmus of Rotterdam, the great humanist.

SECTION C

The Background of the Reformation

119. Tyndale's Criticism of the English Church, 1528

England was generally orthodox on the eve of Henry VIII's break with Rome. But some practices of the church had been criticized since the time of Wycliffe and even earlier, and in the 1520's Lutheranism made inroads in some parts of the country. The following criticism was written by William Tyndale, one of the earliest leaders of the English reformation movement, best known for his translation of the Bible into English.

Mark well how many parsonages or vicarages are there in the realm, which at the least have a plow land[10] apiece. Then note the lands of bishops, abbots, priors, nuns, knights of St. John,[11] cathedral churches, colleges, chantries and free-chapels. For though the house fall in decay, and the ordinance of the founder be lost, yet will not they lose the lands. What cometh once in may never more out. They make a free-chapel[12] of it, so that he which enjoyeth it shall do nought therefor. Besides all this, how many chaplains do gentlemen find at their own cost, in their houses? How many sing for souls, by testaments? Then the proving of testaments,[13] the prizing[14] of goods, the Bishop of Canterbury's prerogative; is that not much through the realm in a year? Four offering days, and privy tithes. There is no servant, but that he shall pay somewhat of his wages. None shall receive the body of Christ at Easter, be he never so poor a beggar, or never so young a lad or maid, but they must pay somewhat for it.

Then mortuaries[15] for forgotten tithes, as they say. And yet what parson or vicar is there that will forget to have a pigeon-house, to peck up somewhat both at sowing time and harvest, when corn is ripe? They

---

SOURCE: William Tyndale, *The Obedience of a Christian Man* (London: Cambridge University Press, Parker Society, 1848), 236–38.

---

10 The amount of land that can be tilled with one plow.
11 The Knights Hospitallers, a military order.
12 A chapel not subject to the usual ecclesiastical authorities.
13 Probate of wills.
14 Appraising.
15 Customary gifts due the priest upon the death of a parishioner.

will forget nothing. No man shall die in their debt: or if any man do, he shall pay it when he is dead. They will lose nothing. Why? It is God's; it is not theirs. It is St. Hubert's rents, St. Alban's lands, St. Edmond's right, St. Peter's patrimony, say they, and none of ours. Item, if a man die in another man's parish, besides that he must pay at home a mortuary for forgotten tithes, he must there pay also the best that he there hath, whether it be an horse of twenty pound, or how good soever he be . . . It is much, verily, for so little pains-taking in confession and in ministering the sacraments. Then bead rolls.[16] Item chrism,[17] churching,[18] banns, wedding, offering at weddings, offering at buryings, offering to images, offering of wax and lights, which come to their vantage; besides the superstitious waste of wax in torches and tapers throughout the land.

Then brotherhood and pardoners.[19] What get they also by confessions? Yea, and many enjoin penance, to give a certain [sum] for to have so many masses said, and desire to provide a chaplain themselves: soul-masses, dirges, month-minds,[20] year-minds,[21] All-Souls-day, and trentals.[22] The mother church and the high altar must have somewhat in every testament. Offerings at priests' first masses. Item, no man is professed of whatsoever religion it be, but he must bring somewhat. The hallowing, or rather conjuring, of churches, chapels, altars, super-altars, chalice, vestments and bells. Then books, bell, candlestick, organs, chalice, vestments, copes, altar-cloths, surplices, towels, basins, ewers, ship,[23] censer and all manner ornament must be found them freely; they will not give a mite thereunto. Last of all, what swarms of begging friars are there! The parson sheareth, the vicar shaveth, the parish priest polleth,[24] the friar scrapeth, and the pardoner pareth; we lack but a butcher to pull off the skin.

120. The Scriptures in English

Prominent in the program of the reformers was the translation of Holy Writ from Latin into the language of the people. We are reminded of

---

SOURCE: William Tyndale, *The Obedience of a Christian Man* (London: Cambridge University Press, Parker Society, 1848), 145–46.

---

16 A list of persons (or objects) for whom prayers were said.
17 Baptism.
18 A ceremony marking the recovery of a woman from childbirth.
19 Persons who sell papal pardons and indulgences.
20 Prayers for a deceased person during the month following his death.
21 Services on the anniversary of a person's death.
22 A service of thirty masses for a deceased person.
23 A vessel for containing incense.
24 Cuts short.

Wycliffe's activities along this line. William Tyndale urged the practice in his *Obedience of a Christian Man,* published in 1528, from which these excerpts are taken.

. . . Moses saith (Deuteronomy 6), "Hear, Israel, let these words which I command thee this day stick fast in thine heart, and whet them on thy children, and talk of them as thou sittest in thine house and as thou walkest by the way and when thou liest down and when thou riseth up, and bind them for a token of thine hand, and let them be a remembrance between thine eyes, and write them on the posts and gates of thine house."[25] This was commanded generally unto all men. How cometh that God's word pertaineth less unto us than unto them? Yea, how cometh it that our Moseses forbid us and command us the contrary, and threat us if we do, and will not that we once speak of God's word? How can we whet God's word (that is, put in practice, use and exercise) upon our children and household, when we are violently kept from it and know it not? . . .

They will say haply, "the Scripture requireth a pure mind and a quiet mind. And therefore the layman, because he is altogether cumbered with worldly business, cannot understand them." If that be the case, then it is a plain case that our prelates understand not the Scriptures themselves. For no layman is so tangled with worldly business as they are. The great things of the world are ministered by them. . . .

"If the Scripture were in the mother tongue," they will say, "then would the lay people understand it every man after his own ways." Wherefore serveth the curate but to teach them the right way? Wherefore were the holidays made but that the people should come and learn? Are ye not abominable schoolmasters in that ye take so great wages, if ye will not teach? If ye would teach, how could ye do it so well and with so great profit as when the lay people have the Scripture before them in their mother tongue? For then should they see, by the order of the text, whether thou juggledest or not. . . .

But alas, the curates themselves (for the most part) wot no more what the New or the Old Testament meaneth than do the Turks. Neither know they of any more than that they read at mass, matins and evensong, which yet they understand not. . . . If they will not let the layman have the word of God in his mother tongue, yet let the priests have it, which, for a great part of them, do understand no Latin at all; but sing and say and patter all day with the lips only that which the heart understandeth not.

[25] A digest of Deuteronomy 6: 4–9.

121. Henry VIII Attacks Luther

In the 1520's, far from manifesting any sympathy with the reformers, Henry VIII maintained an orthodox position in religion, even to the extent of producing a treatise against Martin Luther, *Assertio Septem Sacramentorum* (Defense of the Seven Sacraments). This work earned for him, at the hands of the Pope, the title of Defender of the Faith.

We have in this little book, gentle reader, clearly demonstrated, I hope, how absurdly and impiously Luther has handled the holy sacraments. For though we have not touched all things contained in his book, yet so far as was necessary to defend the sacraments (which was our only design), I suppose I have treated, though not so sufficiently as might have been done, yet more than is even necessary. . . .

But that others may understand how false and wicked his doctrine is, lest they might be so far deceived as to have a good opinion of him, I doubt not but in all parts there are very learned men . . . who have much more clearly discovered the same, than can be shown by me. And if there be any who desire to know this strange work of his, I think I have sufficiently made it apparent to them. For seeing by what has been said, it is evident to all men what sacrilegious opinions he has of the sacrament of our Lord's Body, from which the sanctity of all the other sacraments flow: who would have doubted, if I had said nothing else, how unworthily, without scruple, he treats all the rest of the sacraments? Which, as you have seen, he has handled in such sort that he abolishes and destroys them all, except Baptism alone. . . .

What everybody believes, he alone by his vain reason laughs at, denouncing himself to admit nothing but clear and evident Scriptures. And these, too, if alleged by any against him, he either evades by some private exposition of his own, or else denies them to belong to their own authors. None of the Doctors are so ancient, none so holy, none of so great authority in treating of Holy Writ, but this new doctor, this little saint, this man of learning, rejects with great authority.

Seeing, therefore, he despiseth all men and believes none, he ought not to take it ill if everybody discredit him again. I am so far from holding any further dispute with him that I almost repent myself of what I have already argued against him. For what avails it to dispute against one who disagrees with everyone, even with himself? Who affirms in one place what he denies in another, denying what he presently affirms? Who, if you object faith, combats by reason; if you touch him with

---

SOURCE: *Assertio Septem Sacramentorum* (London: N. Thompson, 1688), 112–14.

reason, pretends faith? If you allege philosophers, he flies to Scripture; if you propound Scripture, he trifles with sophistry. Who is ashamed of nothing, fears none, and thinks himself under no law. Who contemns the ancient Doctors of the church, and derides the new ones in the highest degree; loads with reproaches the Chief Bishop of the church. Finally, he so undervalues the customs, doctrine, manners, laws, decrees and faith of the church (yea, the whole church itself) that he almost denies there is any such thing as a church, except perhaps such a one as himself makes up of two or three heretics, of whom himself is chief. . . .

122. Wolsey on the Divorce, 1527

Henry VIII began action for a divorce from Queen Katherine in 1527. More precisely, he sought an annulment of the marriage, on technical grounds. His real motives, however, centered around his concern for the succession (his marriage to Katherine having produced no male heir) and his desire to marry Anne Boleyn. The following comment by Cardinal Wolsey dates from December, 1527.

I have told you already how the King, partly by his assiduous study and learning and partly by conference with theologians, has found his conscience somewhat burdened with his present marriage; and out of regard to the quiet of his soul, and next to the security of his succession and the great mischiefs likely to arise, he considers it would be offensive to God and man if he were to persist in it, and with great remorse of conscience has now for a long time felt that he is living under the offence of the Almighty, whom in all his efforts and his actions he always sets before him. He has made diligent inquiry whether the dispensation granted for himself and the Queen (as his brother's wife) is valid and sufficient, and he is told that it is not. The bull of dispensation is founded on certain false suggestions, as that his Majesty desired the marriage for the good understanding between Henry VII, Ferdinand and Isabella; whereas there was no suspicion of any misunderstanding between them. And secondly, he never assented or knew anything of this bull, nor wished for the marriage. On these grounds it is judged inefficacious. Next, when the King reached the age of fourteen, the contract was revoked and Henry VII objected to the marriage. To this the King attributes the death of all his male children, and dreads the heavy wrath of God if he persists. Notwithstanding his scruples of conscience, he is resolved to apply for his remedy to the Holy See, trusting that, out of his

---

SOURCE: J. S. Brewer (ed.), *Letters and Papers Foreign and Domestic of the Reign of Henry VIII* (London: Her Majesty's Stationery Office, 1872), *IV*, pt. ii, no. 3641.

consideration of his services to the church, the Pope will not refuse to remove this scruple out of the King's mind, and discover a method whereby he may take another wife, and, God willing, have male children. . . .

SECTION D

The Break with Rome

123. The Reformation Parliament Summoned, 1529

In 1529 the royal divorce case was heard by Cardinal Wolsey and Cardinal Campeggio, an Italian, acting jointly as papal legates. No decision was reached; Pope Clement sought to delay proceedings and finally summoned the case to Rome. At this juncture, Henry turned to Parliament. Before it was dissolved, over six years later, it had enacted legislation ending the papal jurisdiction in England. The Imperial ambassador, Chapuys, had these comments for the Emperor Charles a few weeks before this Parliament assembled. It is noteworthy that, although several years passed before Parliament sanctioned Henry's headship of the English Church, Chapuys was conscious of this possibility.

It is reported that the real cause of this Parliament having been convoked for the 2nd of November is, independently of others specified in my dispatch of the 4th of September, to investigate the conduct and examine the accounts of all those functionaries who have been connected with the finances of the country. Others add that a motion will be made to abolish the legatine office in England, and prevent the Pope from appointing or sending in future legates to this country. Those that think so may not be far from the truth, for I now recollect that at the last sitting of the Legates for the purpose of proroguing the case until the 2nd of October, the Duke of Suffolk[26] got into a great passion and began

SOURCE: Pascal de Gayangos (ed.), *Calendar of State Papers, Spanish* (London: Her Majesty's Stationery Office, 1879), *IV*, pt. i, no. 160.

[26] Charles Brandon, Duke of Suffolk, the King's brother-in-law, who supported Henry's efforts to gain the divorce.

to swear, and say within hearing of the King himself, of the cardinals, and of all those who had come to that piteous ceremony in order to hear whether the sentence was in favor of the Queen or against her: "I see now the truth of what I have heard many people say; never at any time did a papal legate do anything to the profit of England; they have always been, and will hereafter be, a calamity and a sore to this country."

I need scarcely observe that if these sentiments of the Duke gain ground with the King and the people of this country, there will be a door wide open for the Lutheran heresy to creep into England, which is the very identical threat made by the English ambassador at Rome when the Pope was pleased to grant the advocation,[27] as I have informed your Majesty in a previous dispatch. I firmly believe that if they had nothing to fear but the Pope's excommunication and malediction, there are innumerable people in this country who would follow the Duke's advice, and make of the King and ordinary prelates as many popes. All this for the sole purpose of having the divorce case tried in England, notwithstanding the Holy Father's inhibition, and not so much perhaps for the ill-will they bear towards ecclesiastics in general, but principally on account of their property, which they covet and wish to seize. It is to be hoped, however, that fear of your Majesty, if no other consideration, will defeat such wicked plans. . . .

124. The Submission of the Clergy, 1532

By 1532 Henry VIII had forced the clergy of the English Church to agree not to pass any new canons (church laws) without the royal consent, and to approve of a revision of existing canons by the King and a royal commission. He intimidated the clergy by invoking the penalties of praemunire, with reference to their recognition of the legatine authority of Cardinal Wolsey, ignoring the fact that he himself had sanctioned it.

We your most humble servants, daily orators and bedesmen[28] of your clergy of England, having our special trust and confidence in your most excellent wisdom, your princely goodness and fervent zeal to the promotion of God's honor and Christian religion, and also in your learning far exceeding, in our judgment, the learning of all other kings and princes

---

SOURCE: Henry Gee and William J. Hardy, *Documents Illustrative of English Church History* (London: Macmillan, 1921), 176–78.

---

27 The summoning of the action.
28 Men engaged in praying.

that we have read of, and doubting nothing but that the same shall still continue and daily increase in your Majesty:

First, [we] do offer and promise *in verbo sacerdotii*[29] here unto your Highness . . . that we will never from henceforth enact, put in use, promulge or execute any new canons or constitutions provincial, or any other new ordinance . . . in our Convocation . . . unless your Highness by your royal assent shall license us to assemble our Convocation, and to make . . . such constitutions and ordinances as shall be made in the same; and thereto give your royal assent and authority.

Secondly, that whereas divers of the constitutions, ordinances, and canons, provincial or synodal, which have been heretofore enacted, be thought to be not only much prejudicial to your prerogative royal, but also overmuch onerous to your Highness' subjects, your clergy aforesaid is contented, if it may stand so with your Highness' pleasure, that it be committed to the examination and judgment of your Grace, and of thirty-two persons, whereof sixteen to be of the upper and nether house of the temporalty, and the other sixteen of the clergy, all to be chosen and appointed by your most noble Grace. So that, finally, whichsoever of the said constitutions, ordinances, or canons, provincial or synodal, shall be thought and determined by your Grace and by the most part of the said thirty-two persons not to stand with God's laws and the laws of the realm, the same to be abrogated and taken away by your Grace and the clergy; and such of them as shall be seen by your Grace, and by the most part of the said thirty-two persons, to stand with God's laws and the laws of your realm, to stand in full strength and power, your Grace's most royal assent and authority once impetrate[30] and fully given to the same.

125. The Prohibition of Appeals to Rome, 1533

As might be expected, Parliament enacted, in 1533, legislation prohibiting appeals to Rome. The preamble to this measure deserves attention, as epitomizing the political philosophy of those, like Thomas Cromwell, who were determined to end the papal authority in England.

Where, by divers sundry old authentic histories and chronicles, it is manifestly declared and expressed that this realm of England is an

---

SOURCE: *Statutes of the Realm* (London: Her Majesty's Stationery Office, 1810–28), *III*, 427.

---

[29] On the word of a priest.
[30] Obtained by entreaty.

empire, and so hath been accepted in the world, governed by one supreme head and king having the dignity and royal estate of the imperial crown of the same, unto whom a body politic, compact[31] of all sorts and degrees of people, divided in terms and by names of spirituality and temporalty, be bounden and owe to bear next to God a natural and humble obedience (he being also institute and furnished by the goodness and sufferance of Almighty God with plenary, whole, and entire power, pre-eminence, authority, prerogative, and jurisdiction to render and yield justice and final determination to all manner of folk residents or subjects within this his realm, in all causes, matters, debates, and contentions happening to occur, insurge, or begin within the limits thereof, without restraint or provocation to any foreign princes or potentates of the world . . .); and whereas the King, his most noble progenitors and the nobility and commons of this said realm, at divers and sundry parliaments as well in the time of King Edward I, Edward III, Richard II, Henry IV, and other noble kings of this realm, made sundry ordinances, laws, statutes, and provisions for the entire and sure conservation of the prerogatives, liberties, and pre-eminences of the said imperial crown of this realm, and of the jurisdictions spiritual and temporal of the same, to keep it from the annoyance as well of the see of Rome as from the authority of other foreign potentates attempting the diminution or violation thereof, as often and from time to time as any such annoyance or attempt might be known or espied[32]. . .

126. The Act of Supremacy, 1534

In 1534 Henry VIII's position as supreme head of the church in England was recognized by statute, though the King did not formally accept the title until the following year. Repealed by Queen Mary, the act was restated under Elizabeth.

Albeit the King's Majesty justly and rightfully is and ought to be the supreme head of the Church of England, and so is recognized by the clergy of this realm in their Convocations, yet nevertheless for corroboration and confirmation thereof, and for increase of virtue in Christ's religion within this realm of England, and to repress and extirp all errors,

---

SOURCE: *Statutes of the Realm* (London: Her Majesty's Stationery Office, 1810–28), *III*, 492.

---

31 Composed.
32 Here follows the prohibitory section.

heresies and other enormities and abuses heretofore used in the same:

Be it enacted by authority of this present Parliament that the King, our sovereign lord, his heirs and successors, kings of this realm, shall be taken, accepted and reputed the only supreme head in earth of the Church of England, called *Anglicana Ecclesia;* and shall have and enjoy, annexed and united to the imperial crown of this realm, as well the title and style thereof as all honors, dignities, pre-eminences, jurisdictions, privileges, authorities, immunities, profits and commodities to the said dignity of supreme head of the same Church belonging and appertaining; and that our said sovereign lord, his heirs and successors, kings of this realm, shall have full power and authority from time to time to visit, repress, redress, reform, order, correct, restrain and amend all such errors, heresies, abuses, offenses, contempts and enormities, whatsoever they be, which by any manner, spiritual authority or jurisdiction ought or may lawfully be reformed, repressed, ordered, redressed, corrected, restrained or amended, most to the pleasure of Almighty God, the increase of virtue in Christ's religion, and for the conservation of the peace, unity and tranquillity of this realm, any usage, custom, foreign law, foreign authority, prescription or any other thing or things to the contrary hereof notwithstanding.

SECTION E

The Henrician Church

127. The Practice of Religion: Novel and Traditional Precepts

With the papal authority removed, it was necessary for the state to issue regulations regarding religious belief and practice. The idea that individuals might work out their own salvation was acceptable to very few: most men believed that religious uniformity was not only desirable but essential. We find both new and old currents of thought influencing these arrangements: the former in the royal injunctions of 1538, the latter in the Six Articles of the following year. Clearly, religion had entered into the state of flux which has characterized its modern position.

SOURCES: David Wilkins, *Concilia Magnae Britannae* (London: R. Gosling, 1737), *III*, 815–16.

Statutes of the Realm (London: Her Majesty's Stationery Office, 1810–28), *III*, 739–40.

The Royal Injunctions of 1538 – In the name of God, amen. By the authority and commission of the most excellent Prince Henry, by the grace of God King of England and of France, Defender of the Faith, Lord of Ireland, and in earth Supreme Head under Christ of the Church of England, I, Thomas, Lord Cromwell, Lord Privy Seal, vicegerent to the King's said Highness for all his jurisdictions ecclesiastical within this realm, do . . . give and exhibit unto you . . . these injunctions following, to be kept, observed and fulfilled upon the pains hereafter declared.

First, that you shall truly observe and keep all and singular the King's Highness's injunctions given unto you heretofore in my name by his Grace's authority. . . .

Item, that you shall provide . . . one book of the whole Bible of the largest volume, in English, and the same set up in some convenient place within the . . . church that you have cure of, whereas your parishioners may most commodiously resort to the same and read it. . . .

Item, that you discourage no man privily or apertly[33] from the reading or hearing of the said Bible, but shall expressly provoke, stir and exhort every person to read the same, as that which is the very lively word of God that every Christian man is bound to embrace, believe and follow, if he look to be saved; admonishing them, nevertheless, to avoid all contention and altercation therein, and to use an honest sobriety in the inquisition of the true sense of the same, and refer the explication of obscure places to men of higher judgment in Scripture.

Item, that you shall every Sunday and holy day through the year openly and plainly recite to your parishioners twice or thrice together, or oftener if need require, one particle or sentence of the Paternoster or Creed, in English, to the intent they may learn the same by heart, and so from day to day to give them one like lesson or sentence of the same, till they have learned the whole Paternoster and Creed in English, by rote; and as they be taught every sentence of the same . . . you shall expound and declare the understanding of the same unto them, exhorting all parents and householders to teach their children and servants the same, . . . and that done, you shall declare unto them the Ten Commandments, one by one, every Sunday and holy day, till they be likewise perfect in the same. . . .

Item, that you shall make, or cause to be made . . . one sermon every quarter of the year at the least, wherein you shall purely and sincerely declare the very Gospel of Christ, and in the same exhort your hearers

[33] Openly.

to the works of charity, mercy and faith specially prescribed and commanded in Scripture, and not to repose their trust or affiance[34] in any other works devised by men's phantasies beside Scripture; as in wandering to pilgrimages, offering of money, candles or tapers to images or relics, or kissing or licking the same, saying over a number of beads, not understood or minded on, or in such-like superstition, for the doing thereof you not only have no promise of reward in Scripture, but contrariwise great threats and maledictions of God, as things tending to idolatry and superstition, which of all other offenses God Almighty does most detest and abhor, for that the same diminishes most his honor and glory.

Item, that such feigned images as you know in any of your cures[35] to be so abused with pilgrimages or offerings of anything made thereunto, you shall for avoiding that most detestable offence of idolatry forthwith take down . . . and shall suffer from henceforth no candles, tapers, or images of wax to be set afore any image or picture, but only the light that commonly goeth across the church by the rood loft, the light before the Sacrament of the altar, and the light about the sepulcher,[36] which for the adorning of the church and divine service you shall suffer to remain. . . .

Item, if you do or shall know any man within your parish or elsewhere that is a letter[37] of the word of God to be read in English, or sincerely preached, or of the execution of these Injunctions, or a fautor[38] of the Bishop of Rome's pretensed power, now by the law of this realm justly rejected and extirpated, you shall detect and present the same to the King's Highness, or his honorable Council, or to his vicegerent aforesaid, or the justice of peace next adjoining. . . .

An Act Abolishing Diversity in Opinions (the "Six Articles") — Where[as] the King's most excellent Majesty is by God's law supreme head immediately under Him of this whole church and congregation of England, intending the conservation of the same . . . in a true, sincere and uniform doctrine . . . , and . . . hath therefore caused and commanded that his most high Court of Parliament . . . to be at this time summoned, and also a synod and convocation of all the archbishops, bishops, and other learned men of the clergy of this his realm to be in like manner assembled; and forasmuch as in the said Parliament, synod and convocation there were certain articles, matters and questions proponed[39] and set forth touching Christian religion . . . :

[34] Faith.
[35] Curacies.
[36] A recess in the church structure, used for Good Friday and Easter rites.
[37] Hinderer.
[38] Favorer.
[39] Proposed.

Whereupon, after a great and long deliberate and advised disputation and consultation had and made concerning the said articles, . . . it was and is finally resolved, accorded and agreed in manner and form following, that is to say:

First, that in the most blessed sacrament of the altar, by the strength and efficacy of Christ's mighty word, it being spoken by the priest, is present really, under the form of bread and wine, the natural body and blood of our Saviour Jesu Christ, conceived of the Virgin Mary, and that after the consecration there remaineth no substance of bread and wine, nor any other substance but the substance of Christ, God and man; secondly, that communion in both kinds is not necessary *ad salutem*[40] by the law of God to all persons, and that it is to be believed and not doubted of but that in the flesh under form of bread is the very blood, and with the blood under form of wine is the very flesh, as well apart as though they were both together; thirdly, that priests, after the order of priesthood received as afore, may not marry by the law of God; fourthly, that vows of chastity or widowhood by man or woman made to God advisedly ought to be observed by the law of God, and that it exempteth them from other liberties of Christian people which without that they might enjoy; fifthly, that it is meet and necessary that private masses be continued and admitted in this the King's English church and congregation, as whereby good Christian people ordering themselves accordingly do receive both godly and goodly consolations and benefits, and it is agreeable also to God's law; sixthly, that auricular[41] confession is expedient and necessary to be retained and continued, used and frequented in the church of God. . . .

128. The Investigation of the Monasteries

The way was paved for the dissolution of the monastic institutions by reports of commissioners, dispatched by the government to investigate the conditions prevailing there. It seems clear that these agents were expected to find fault; hence their reports must be used with caution. The following example, presented by John Ap Rice to Thomas Cromwell, Henry VIII's chief administrative aide of the time, concerns the great monastic house at Bury St. Edmunds.

SOURCE: Thomas Wright (ed.), *Letters Relating to the Suppression of the Monasteries* (London: Royal Historical Society, 1843), 85–86.

---

40 For salvation.
41 Privately heard.

Please it your mastership: Forasmuch as I suppose you shall have suit made unto you touching Bury ere we return, I thought convenient to advise you of our proceedings there, and also of the comports[42] of the same. As for the abbot, we found nothing suspect as touching his living, but it was detected that he lay much forth in his granges,[43] that he delighted much in playing at dice and cards, and therein spent much money, and in building for his pleasure. He did not preach openly. Also that he converted divers farms[44] into copyholds, whereof poor men doth complain. Also he seemeth to be addict to the maintaining of such superstitious ceremonies as hath been used heretofore.

As touching the convent, we could get little or no reports among them, although we did use much diligence in our examination, and thereby, with some other arguments gathered of their examinations, I firmly believe and suppose that they had confedered[45] and compacted before our coming that they should disclose nothing. And yet it is confessed and proved that there was here such frequence of women coming and resorting to this monastery as to no place more. Amongst the relics we found much vanity and superstition, as the coals that St. Lawrence was toasted withal, the paring of St. Edmund's nails, St. Thomas of Canterbury's pen-knife and his boots, and divers skulls for the headache, pieces of the holy cross able to make a whole cross of, other relics for rain and certain other superstitious usages, for avoiding of weeds growing in corn, with such other. Here depart of them that be under age upon an eight,[46] and of them that be above age upon a five would depart if they might, and they be of the best sort in the house, and of best learning and judgment. The whole number of the convent before we came was sixty, saving one, beside three that were at Oxford. Of Ely I have written to your mastership by my fellow, Richard a Lee. And thus Almighty God have you in his tuition. From Bury, 5 November.

Your servant most bounden,
John Ap Rice

129. The Dissolution of the Lesser Monasteries, 1536

In 1536 Parliament passed an act abolishing the lesser monasteries (those having incomes of less than £200 a year), and conferring their

SOURCE: *Statutes of the Realm* (London: Her Majesty's Stationery Office, 1810–28), *III*, 575.

[42] Behavior.
[43] Outlying farm properties attached to monasteries.
[44] Rents.
[45] Confederated, joined together.
[46] About eight.

property on the King. Thus began the destruction of these institutions, save for a few dissolved under Cardinal Wolsey. The process was completed when a similar statute ordered the dissolution of the greater houses in 1539.

Forasmuch as manifest sin, vicious, carnal and abominable living is daily used and committed among the little and small abbeys, priories and other religious houses of monks, canons and nuns, where the congregation of such religious persons is under the number of twelve persons, whereby the governors of such religious houses, and their convent, spoil, destroy, consume and utterly waste as well their churches, monasteries, priories, principal houses, farms, granges, lands, tenements and hereditaments as the ornaments of their churches and their goods and chattels, to the high displeasure of Almighty God, slander of good religion, and to the great infamy of the King's Highness and the realm, if redress should not be had thereof:

And albeit that many continual visitations hath been heretofore had, by the space of 200 years and more, for an honest and charitable reformation of such unthrifty carnal and abominable living, yet nevertheless little or none amendment hath been hitherto had, but their vicious living shamelessly increaseth and augmenteth, and by a cursed custom so rooted and infested, that a great multitude of the religious persons in such small houses do rather choose to rove abroad in apostasy than to conform themselves to the observation of good religion, so that without such small houses be utterly suppressed and the religious persons therein committed to great and honorable monasteries of religion in this realm, where they may be compelled to live religiously by reformation of their lives, there cannot else be no reformation in this behalf:

In consideration whereof the King's most royal Majesty, being supreme head on earth, under God, of the Church of England, daily finding and devising the increase, advancement and exaltation of true doctrine and virtue in the said Church, to the only glory and honor of God and the total extirping and destruction of vice and sin, having knowledge that the premises be true, . . . considering also that divers and great solemn monasteries of this realm, wherein (thanks be to God) religion is right well kept and observed, be destitute of such full numbers of religious persons as they ought and may keep, hath thought good that a plain declaration should be made of the premises, as well to the Lords spiritual and temporal as to other his loving subjects, the Commons, in this present Parliament assembled:

Whereupon the said Lords and Commons, by a great deliberation, finally be resolved that it is and shall be much more to the pleasure of Almighty God, and for the honor of this his realm, that the possessions of such small religious houses . . . should be used and converted to better uses, and the unthrifty religious persons . . . be compelled to reform their lives; and thereupon most humbly desire the King's Highness that it may be enacted by authority of this present Parliament that his Majesty shall have and enjoy to him and his heirs forever all and singular such monasteries, priories and other religious houses of monks, canons and nuns, of what kinds of diversities of habits, rules or orders soever they be called or named, which have not in lands and tenements, rents, tithes, portions and other hereditaments above the clear yearly values of £200.

SECTION F

Religious Change under Edward VI and Mary

130. Manifestations of Religious Change

As the current of Protestantism ran stronger, it was accompanied by the destruction of images, attacks on the Mass, and the increased use of English in religious services, as can be seen from these excerpts from the Greyfriars Chronicle (1547–1548), kept by the Franciscans.

Item, . . . in September began the King's visitation at Paul's, and all images pulled down; and the 9th day of the same month the said visitation was at St. Bride's, and after that in divers other parish churches; and so all images pulled down through all England at that time, and all churches new white-limed,[47] with the Commandments written on the walls. . . .

Item, at this time was pulled up all the tombs, great stones, all the altars, with the stalls and walls of the choir and altars in the church that was sometime the Greyfriars', and sold, and the choir made smaller. . . .

Item, the 17th day of the same month at night was pulled down the rood[48] in Paul's, with Mary and John,[49] with all the images in the church,

---

SOURCE: John G. Nichols (ed.), *Chronicle of the Grey Friars of London* (London: Royal Historical Society, 1852), 54–55.

---

[47] Whitewashed.
[48] A large cross or crucifix at the entrance to the chancel.
[49] I.e., statues of the Virgin Mary and St. John.

and two of the men that labored at it was slain and divers others sore hurt. Item, also at that time was pulled down through all the King's dominion in every church all roods with all images, and every preacher preached in their sermons against all images. . . . Also this same time was much speaking against the sacrament of the altar . . . ; and then was made a proclamation against such sayers,[50] and yet both the preachers and others spake against it, and so continued; and at Easter following there began the communion, and confession but of those that would,[51] as the book doth specifieth. And at this time was much preaching against the Mass, and the sacrament of the altar pulled down in divers places through the realm. Item, after Easter began the service in English (at Paul's at the commandment of the Dean at the time, William May), and also in divers other parish churches. . . .

131. The Second Act of Uniformity, 1552

The religious conservatism of Henry VIII had acted as a restraint upon the introduction of Protestant beliefs and practices. But with the accession of his son, Edward VI, they made marked headway. Edward's first Act of Uniformity (1549), introducing a Book of Common Prayer in the English tongue, was moderate enough; but the second Act (1552), with a revised Prayer Book, was much more definitely Protestant in tone. It will be noted that all subjects throughout the realm were called upon to participate in the new service: there was no provision for diversity of opinion.

Be it enacted . . . that, from and after the Feast of All Saints next coming, all and every person and persons inhabiting within this realm or any other the King's Majesty's dominions shall diligently and faithfully, having no lawful or reasonable excuse to be absent, endeavor themselves to resort to their parish church or chapel accustomed, or upon reasonable let thereof to some usual place where common prayer and such service of God shall be used in such time of let, upon every Sunday and other days ordained and used to be kept as holy days, and then and there to bishops, bishops and other ordinaries[52] that they shall endeavor them-

---

SOURCE: *Statutes of the Realm* (London: Her Majesty's Stationery Office, 1810–28), *IV*, 130.

---

[50] Those who spoke thus.
[51] I.e., the communion replaced the mass, and confession was voluntary.
[52] Ecclesiastical officials.

selves to the uttermost of their knowledge that the due and true execution hereof may be had throughout their dioceses and charges . . .
abide orderly and soberly during the time of common prayer, preachings or other service of God there to be used and ministered, upon pain of punishment by the censures of the church.

And for the due execution hereof the King's most excellent Majesty, the Lords temporal and all the Commons in this present Parliament assembled do in God's name earnestly require and charge all the arch-

And because there hath arisen in the use and exercise of the aforesaid common service in the church heretofore set forth divers doubts for the fashion and manner of the ministration of the same, . . . therefore, as well for the more plain and manifest explanation hereof as for the more perfection of the said order of common service, in some places where it is necessary to make the same prayers and fashion of service more earnest and fit to stir Christian people to the true honoring of Almighty God, the King's most excellent Majesty, with the assent of the Lords and Commons in this present Parliament assembled . . . , hath caused the aforesaid order of common service entitled *The Book of Common Prayer* to be faithfully and godly perused, explained and made fully perfect; and by the foresaid authority hath annexed and joined it, so explained and perfected, to this present statute, adding also a form and manner of making and consecrating archbishops, bishops, priests and deacons, to be of like force, authority and value as the . . . aforesaid book entitled The Book of Common Prayer was before, and to be accepted . . . in like sort and manner . . . as by the Act of Parliament made in the second year of the King's Majesty's reign was ordained. . . .

132. The Reunion with Rome under Mary

The accession of Mary in 1553 meant a reversal of the religious policies of the past twenty years. Mary, a staunch Catholic, worked tirelessly to effect a complete restoration of the old faith, and was successful save in that she was unable to re-endow the church with the property it had lost in two reigns. Like her father and brother she worked through Parliament. The second Statute of Repeal shows both the magnitude and the limitations of her accomplishment.

SOURCE: *Statutes of the Realm* (London: Her Majesty's Stationery Office, 1810–28), IV, 246–48.

Whereas, since the twentieth year[53] of King Henry VIII of famous memory, father unto your Majesty, . . . much false and erroneous doctrine hath been taught, preached and written . . . , by reason whereof as well the spiritualty as the temporalty of your Highness's realms and dominions have swerved from the obedience of the See Apostolic and declined from the unity of Christ's church, and so have continued until . . . , your Majesty being . . . raised up by God and set in the seat royal over us . . . , the Pope's Holiness and the See Apostolic sent hither . . . the most reverent father in God, the lord Cardinal Pole, legate *de latere*,[54] to call us home again into the right way . . . ; and we . . . , seeing by the goodness of God our own errors, have acknowledged the same unto the said most reverend father, and by him have been . . . received and embraced into the unity and bosom of Christ's church . . . , upon our humble submission and promise . . . to repeal and abrogate such acts and statutes as had been made in Parliament since the said twentieth year of the said King Henry VIII against the supremacy of the See Apostolic . . . [therefore, such legislation is repealed].

And finally, where certain acts and statutes have been made in the time of the late schism concerning the lands and hereditaments of archbishoprics and bishoprics, the suppression and dissolution of monasteries, abbeys, priories, chantries, colleges, and all other the goods and chattels of religious houses; since the which time the right and dominion of certain lands and hereditaments, goods and chattels, belonging to the same be dispersed abroad and come to the hands and possessions of divers and sundry persons who by gift, purchase, exchange and other means, according to the order of the laws and statutes of this realm for the time being, have the same; for the avoiding of all scruples that might grow by any of the occasions aforesaid or by any other ways or means whatsoever, [we ask that] it may please your Majesties to be intercessors and mediators to the said most reverend father . . . that all such causes and quarrels as by pretence of the said schism or by any other occasion or mean whatsoever might be moved, by the Pope's Holiness or See Apostolic or by any other jurisdiction ecclesiastical, may be utterly removed and taken away; so as all persons having sufficient conveyance of the said lands and hereditaments, goods and chattels, . . . may without scruple of conscience enjoy them . . .

53 I.e., since 1529, when the "Reformation Parliament" first met.
54 Ambassador from the Pope.

SECTION G

The Elizabethan Religious Settlement

133. Elizabeth's Ecclesiastical Supremacy, 1559

The revived Catholic ascendancy ended with the death of Mary in 1558. Her successor, Elizabeth I, identified herself with a moderate Protestant position. In the Act of Supremacy she assumed the governorship of the church, after the manner of her father, though not in precisely the same terms. The following oath, prescribed by the Act, was designed to insure official support for this arrangement.

And for the better observation and maintenance of this Act, may it please your Highness that it may be further enacted . . . that all and every archbishop, bishop, and all and every other ecclesiastical person, . . . of what estate, dignity, pre-eminence or degree soever he or they be or shall be, and all and every temporal judge, justice, mayor and other lay or temporal officer and minister, and every other person having your Highness's fee or wages, within this realm or any of your Highness's dominions, shall make, take and receive a corporal oath upon the evangelist,[55] before such person or persons as shall please your Highness, your heirs or successors under the great seal of England to assign and name to accept and to take the same, according to the tenor and effect hereafter following, that is to say:

"I, A. B.,[56] do utterly testify and declare in my conscience that the Queen's Highness is the only supreme governor of this realm, and of all other her Highness's dominions and countries, as well in all spiritual or ecclesiastical things or causes, as temporal, and that no foreign prince, prelate, state or potentate hath or ought to have any jurisdiction, power, superiority, pre-eminence or authority ecclesiastical or spiritual within this realm; and therefore I do utterly renounce and forsake all foreign jurisdictions, powers, superiorities and authorities, and do promise that from henceforth I shall bear faith and true allegiance to the Queen's Highness, her heirs and lawful successors, and to my power shall assist and defend

---

SOURCE: *Statutes of the Realm* (London: Her Majesty's Stationery Office, 1810–28), IV, 352.

---

55 One of the four Gospels.
56 Here the individual stated his name.

all jurisdictions, pre-eminences, privileges and authorities granted or belonging to the Queen's Highness, her heirs or successors, or united or annexed to the imperial crown of this realm. So help me God, and by the contents of this book.". . .

134. Elizabeth's Act of Uniformity, 1559

Elizabeth, drawing upon moderate Protestant belief and Catholic ceremonial, sought to establish a national church which would be acceptable to the majority of her subjects. Uniformity of religious practice continued to be regarded as essential; hence a third Act of Uniformity was obtained from Parliament, restoring the Prayer Book of 1552 (with some revisions) and imposing a fine on those not attending the prescribed services.

Where[as] at the death of our late sovereign lord King Edward VI there remained one uniform order of common service and prayer, and of the administration of sacraments, rites and ceremonies in the Church of England, which was set forth in one book, entitled *The Book of Common Prayer and Administration of Sacraments and Other Rites and Ceremonies in the Church of England,* authorized by Act of Parliament holden in the fifth and sixth years of our said sovereign lord King Edward VI . . . ; the which was repealed and taken away by Act of Parliament in the first year of the reign of our late sovereign lady Queen Mary, to the great decay of the due honor of God, and discomfort to the professors of the truth of Christ's religion:

Be it therefore enacted by the authority of this present Parliament that the said statute of repeal . . . shall be void and of none effect, from and after the Feast of the Nativity of St. John Baptist next coming; and that the said book, with the order of service and of the administration of sacraments, rites and ceremonies, with the alterations and additions therein added and appointed by this statute, shall stand and be, from and after the said Feast . . . , in full force and effect, according to the tenor and effect of this statute; everything in the aforesaid statute of repeal to the contrary notwithstanding.

And further be it enacted by the Queen's Highness, with the assent of the Lords and Commons in this present Parliament assembled, . . . that all and singular ministers in any cathedral or parish church, or other place within this realm of England, Wales, and the marches[57] of the same, or

---

SOURCE: *Statutes of the Realm* (London: Her Majesty's Stationery Office, 1810–28), *IV*, 355–57.

---

[57] Borderlands, frontier regions.

other the Queen's dominions, shall, from and after the Feast of the Nativity of St. John Baptist next coming, be bound to say and use the Matins, Evensong, celebration of the Lord's Supper and administration of each of the sacraments, and all the common and open prayer, in such order and form as is mentioned in the said book so authorized by Parliament in the said fifth and sixth years of the reign of King Edward VI . . .

And that from and after the said Feast . . . all and every person and persons inhabiting within this realm, or any other the Queen's Majesty's dominions, shall diligently and faithfully, having no lawful or reasonable excuse to be absent, endeavor . . . to resort to their parish church or chapel accustomed, . . . upon every Sunday and other days ordained and used to be kept as holy days, and then and there to abide orderly and soberly during the time of the common prayer, preachings or other service of God there to be used and ministered, upon pain of punishment by the censures of the church, and also upon pain that every person so offending shall forfeit for every such offense twelve pence, to be levied by the churchwardens of the parish where such offense shall be done, to the use of the poor of the same parish. . . .

135. Mass or Communion?

The doctrine of transubstantiation,[58] more than any other doctrinal matter, divided Roman Catholics from the various Protestants. In the Elizabethan Prayer Book of 1559 the exact nature of the communion rite was not precisely defined, in the interest of avoiding religious controversy. A juxtaposition of extracts from the communion service in the three sixteenth-century English Prayer Books shows how the Elizabethan liturgy compares with the previous renditions.

[1549] — And when he delivereth the Sacrament of the body of Christ he shall say to everyone these words:

"The body of our Lord Jesus Christ which was given for thee, preserve thy body and soul unto everlasting life."

And the minister delivering the Sacrament of the blood, and giving everyone to drink once and no more, shall say:

SOURCE: *The Book of Common Prayer*, 1549, 1552 and 1559, in J. S. Millward, *Portraits and Documents: Sixteenth Century* (London: Hutchinson Educational, 1961), 64.

---

[58] The change of the bread and the wine into the body and blood of Christ, effected in the Eucharist.

"The blood of our Lord Jesus Christ which was shed for thee, preserve thy body and soul unto everlasting life."

[1552] – And when he delivereth the bread, he shall say:

"Take and eat this in remembrance that Christ died for thee, and feed on him in thy heart by faith, with thanksgiving."

And the minister that delivereth the cup shall say:

"Drink this in remembrance that Christ's blood was shed for thee, and be thankful."

[1559] – And when he delivereth the bread, he shall say:

"The body of our Lord Jesu Christ which was given for thee, preserve thy body and soul into everlasting life, and take, and eat this, in remembrance that Christ died for thee, feed on him in thine heart by faith and thanksgiving."

And the minister that delivereth the cup shall say:

"The blood of our Lord Jesu Christ which was shed for thee, preserve thy body and soul into everlasting life. And drink this in remembrance that Christ's blood was shed for thee, and be thankful."

SECTION H

Catholic Opposition to the Religious Settlement

136. The Excommunication of Queen Elizabeth, 1570

In 1570 the Pope launched a bull of excommunication against Elizabeth, declaring her deposed and absolving Catholic subjects from the bonds of allegiance. The central section of the bull, here omitted, recites various actions of Elizabeth against "the exercise of the true religion."

Pius, Bishop, servant to God's servants, for a future memorial of the matter.

He that reigneth on high, to whom is given all power in heaven and in earth, hath committed his one, holy, catholic and apostolic church, out of

SOURCE: William Camden, *History of . . . Princess Elizabeth* (London: M. Flesher for R. Bentley, 1688), 146.

which there is no salvation, to one alone upon earth, namely to Peter the chief of the apostles, and to Peter's successor, the Bishop of Rome, to be by him governed with plenary authority. Him alone hath he made prince over all people and all kingdoms, to pluck up, destroy, scatter, consume, plant and build, that he may preserve his faithful people (knit together with the band of charity) in the unity of the spirit, and present them spotless and unblamable to their Saviour. In discharge of which function, we, who are by God's goodness called to the government of the aforesaid church, do spare no pains . . .

But the number of the ungodly hath gotten such power that there is now no place in the whole world left which they have not essayed to corrupt with their most wicked doctrines; and amongst others, Elizabeth, the pretended[59] Queen of England, the servant of wickedness, lendeth thereunto her helping hand, with whom, as in a sanctuary, the most pernicious persons have found a refuge. This very woman, having seized on the kingdom and monstrously usurped the place of supreme head of the church in all England and the chief authority and jurisdiction thereof, hath again reduced the said kingdom into a miserable and ruinous condition, which was so lately reclaimed to the Catholic faith and a thriving condition. . . .

We, seeing that impieties and wicked actions are multiplied one upon another, as also that the persecution of the faithful and affliction for religion groweth every day heavier and heavier, through the instigation and by means of the said Elizabeth, and since we understand her heart to be so hardened and obdurate that she hath not only contemned the godly requests and admonitions of Catholic princes concerning her cure and conversion but also hath not so much as suffered the nuncios[60] of this See to cross the seas for this purpose into England, are constrained of necessity to betake ourselves to the weapons of justice against her, being heartily grieved and sorry that we are compelled thus to punish one to whose ancestors the whole state of christendom hath been so much beholden.

Being therefore supported with His authority whose pleasure it was to place us (though unable for so great a burden) in this supreme throne of justice, we do out of the fullness of our apostolic power declare the aforesaid Elizabeth . . . an heretic and a favorer of heretics, and her adherents in the matters aforesaid, to have incurred the sentence of excommunication, and to be cut off from the unity of the body of Christ. And moreover we do declare her to be deprived of her pretended title to the kingdom aforesaid, and of all dominion, dignity and privilege whatsoever; and also the nobility, subjects and people of the said kingdom,

59 Catholics regarded Mary of Scotland as rightful Queen of England.
60 Permanent official representatives of the Pope at a foreign court.

and all others who have in any sort sworn unto her, to be forever absolved from any such oath, and all manner of duty of dominion, allegiance and obedience; and we also do by authority of these presents absolve them, and do deprive the said Elizabeth of her pretended title to the kingdom, and all other things before named. And we do command and charge all and every the noblemen, subjects, people and others aforesaid that they presume not to obey her or her orders, mandates and laws; and those which shall do the contrary we do include them in the like sentence of anathema. . . .

137. The Banishment of Catholic Priests, 1585

The excommunication of Elizabeth, the plots on behalf of Mary Queen of Scots, and the missionary activities of Jesuits and other Catholic priests all contributed to the passage of anti-Catholic laws. Even the attempted conversion of an Englishman to Catholicism was declared to be treasonous. In 1585 an act was passed banishing Jesuits and other Catholic priests from the realm.

Whereas divers persons called or professed Jesuits, seminary priests[61] and other priests, which have been and from time to time are made in the parts beyond the seas by or according to the order and rites of the Romish Church, have of late years come and been sent, and daily do come and are sent, into this realm of England and other the Queen's Majesty's dominions . . . not only to withdraw her Highness' subjects from their due obedience to her Majesty but also to stir up and move sedition, rebellion and open hostility within her Highness' realms and dominions, to the great dangering of the safety of her most royal person and to the utter ruin, desolation and overthrow of the whole realm, if the same be not the sooner by some good means foreseen and prevented:

For reformation whereof be it ordained, established and enacted by the Queen's most excellent Majesty and the Lords spiritual and temporal and the Commons in this present Parliament assembled and by the authority of the same Parliament, that all and every Jesuits, seminary priests and other priests whatsoever, made or ordained out of the realm of England or other her Highness's dominions or within any of her Majesty's realms or dominions by any authority, power or jurisdiction

---

SOURCE: *Statutes of the Realm* (London: Her Majesty's Stationery Office, 1810–28), *IV*, 706.

---

[61] Those educated in foreign seminaries, such as Douai.

derived, challenged or pretended from the see of Rome since the Feast of the Nativity of St. John Baptist in the first year of her Highness' reign, shall within forty days next after the end of this present session of Parliament depart out of this realm of England and out of all other her Highness' realms and dominions, if the wind, weather and passage shall serve for the same; or else so soon after the end of the said forty days as the wind, weather and passage shall so serve.

II. And be it further enacted by the authority aforesaid that it shall not be lawful to or for any Jesuit, seminary priest or other such priest, deacon or any religious or ecclesiastical person whatsoever, being born within this realm or any other her Highness' dominions, and heretofore since the said Feast . . . made, ordained or professed, or hereafter to be made, ordained or professed, by any authority or jurisdiction derived, challenged or pretended from the see of Rome, . . . to come into, be or remain in any part of this realm or any other her Highness' dominions after the end of the same forty days, other than in such special cases and upon such special occasions only and for such time only as is expressed in this Act; and if he do, that then every such offense shall be taken and adjudged to be high treason; and every person so offending shall for his offense be adjudged a traitor, and shall suffer, lose and forfeit as in cases of high treason. And every person which after the end of the same forty days, and after such time of departure as is before limited and appointed, shall wittingly and willingly receive, relieve, comfort, aid or maintain any such Jesuit, seminary priest or other priest, deacon or religious or ecclesiastical person as is aforesaid, being at liberty or out of hold, knowing him to be a Jesuit [etc.], . . . shall also for such offense be adjudged a felon without benefit of clergy,[62] and suffer death, lose and forfeit as in case of one attainted of felony. . . .

138. Governmental Interrogation of Catholics

There was a tendency on the part of the government, after Elizabeth's excommunication, to suspect the loyalty of all Catholics, whether or not they were detected in subversive activities. The following questions were put in 1582 to those suspected of being Catholics, or known to be such.

SOURCE: M. A. Tierney (ed.), *Dodd's Church History of England from the Commencement of the Sixteenth Century to the Revolution of 1688* (London: Charles Dolman, 1839–43), *III*, app. iii, pp. iv–v.

---

62 The right, first confined to bonafide clerics but later extended to laymen as well, to be tried by church courts.

1. Whether the bull[63] of Pius V against the Queen's Majesty be a lawful sentence, and ought to be obeyed by the subjects of England.

2. Whether the Queen's Majesty be a lawful queen and ought to be obeyed by the subjects of England, notwithstanding the bull of Pius V or any other bull or sentence that the Pope hath pronounced, or may pronounce, against her Majesty.

3. Whether the Pope have, or had, power to authorize the Earls of Northumberland, Westmorland, and other of her Majesty's subjects to rebel or take arms against her Majesty;[64] or to authorize Dr. Sanders or others to invade Ireland,[65] or any other her dominions, and to bear arms against her, and whether they did therein lawfully, or not.

4. Whether the Pope hath power to discharge any of her Highness' subjects, or the subjects of any Christian prince, from their allegiance or oath of obedience to her Majesty or to their prince for any cause.

5. Whether the said Dr. Sanders, in his book the *Visible Monarchy of the Church*,[66] and Dr. Bristow, in his book of *Motives*,[67] written in allowance, commendation, and confirmation of the said bull of Pius V, have therein taught, testified, or maintained a truth or a falsehood.

6. If the Pope do by his bull or sentence pronounce her Majesty to be deprived and no lawful queen, and her subjects to be discharged of their allegiance and obedience unto her, and, after, the Pope or any other of his appointment and authority do invade this realm, which part would you take, or which part ought a good subject of England to take?

[63] The bull of excommunication; see p. 201.
[64] The reference is to the Rising of the Northern Earls in 1569, to place Mary Queen of Scots on the English throne and restore the Catholic religion.
[65] An attack under papal sponsorship had been launched in 1578. Dr. Nicholas Sanders was sent to Ireland as papal nuncio to excite rebellion.
[66] *De Visibili Monarchia Ecclesiae* (1571).
[67] Richard Bristow, *A Brief Treatise of Divers Plain and Sure Ways . . . Containing Sundry Worthy Motives unto the Catholic Faith* (Antwerp, 1574).

SECTION I

Protestant Opposition to the Religious Settlement

139. Puritan Demands in Convocation, 1563

In Elizabethan England, the term *Puritan* refers to those Protestants who wished to purify the national church of alleged abuses and defects, whether or not they sought also to alter the organization of the church. In general, they wished to eliminate vestiges of the old religious practice, as is seen from these demands made in Convocation (the assembly of English clerics in each archdiocese) in 1563.

On February the 13th there was a notable matter brought into the lower house; the determination of which matter depended upon a narrow scrutiny of the members. For on the day aforesaid these articles were read, to be approved or rejected:

I. That all the Sundays in the year, and principal feasts of Christ, be kept holy days; and all other holy days to be abrogated.

II. That in all parish churches the minister in common prayer turn his face towards the people; and there distinctly read the divine service appointed, where all the people assembled may hear and be edified.

III. That in ministering the sacrament of baptism, the ceremony of making the cross in the child's forehead may be omitted, as tending to superstition.

IV. That forasmuch as divers communicants are not able to kneel during the time of the communion, for age, sickness, and sundry other infirmities; and some also superstitiously both kneel and knock;[68] that order of kneeling may be left to the discretion of the ordinary within his jurisdiction.

V. That it be sufficient for the minister, in time of saying divine service and ministering of the sacraments, to use a surplice; and that no minister say service, or minister the sacraments, but in a comely garment or habit.

VI. That the use of organs be removed.

Upon this arose a great contest in the house; some saying, they approved of these articles, others not. . . .

SOURCE: John Strype, *Annals of the Reformation* (Oxford: The Clarendon Press, 1812–24, *I*, pt. i, 502–03.

---

[68] Strike one's breast as in penitence.

140. Presbyterian Proposals, 1572

Among the Puritans was a group who favored the establishment of a presbyterian system of church government, in place of the traditional episcopal order. Their position may be seen in *An Admonition to Parliament,* a work published in 1572. Its authors, the clergymen John Field and Thomas Wilcox, suffered imprisonment for their boldness.

Seeing that nothing in this mortal life is more diligently to be sought for and carefully to be looked unto than the restitution of true religion and reformation of God's church, it shall be your parts (dearly beloved), in this present Parliament assembled, as much as in you lieth to promote the same, and to employ your whole labor and study, not only in abandoning all popish remnants both in ceremonies and regiment,[69] but also in bringing in and placing in God's church those things only which the Lord himself in his word commandeth. . . .

May it therefore please your wisdoms to understand, we in England are so far off from having a church rightly reformed according to the prescript of God's word, that as yet we are not come to the outward face of the same. . . . For . . . now by the letters commendatory of some one man, noble or other, tag and rag, learned and unlearned, of the basest sort of people . . . are freely received.[70]. . . Then[71] election was made by the common consent of the whole church; now everyone picketh out for himself some notable good benefice, he obtaineth the next advowson by money or by favor, and so thinketh himself to be sufficiently chosen. . . . Then it was painful, now gainful; then poor and ignominious, now rich and glorious. . . .

Your wisdoms have to remove advowsons, patronages, impropriations and bishops' authority, claiming to themselves thereby right to ordain ministers, and to bring in that old and true election which was accustomed to be made by the congregation. . . . Appoint to every congregation a learned and diligent preacher. Remove homilies, articles, injunctions, a prescript[72] order of service made out of the mass-book. Take away the lordship, the loitering, the pomp, the idleness and livings of bishops. . . .

SOURCE: John Field and Thomas Wilcox, *An Admonition to the Parliament* (Wandsworth: 1572), in Arundel Esdaile, *The Age of Elizabeth, 1547–1603* (London: G. Bell and Sons, 1912), 32–34.

---

69 Regimen, rule.
70 The reference is to the appointment of ministers.
71 In the primitive church.
72 Prescribed, authorized.

To these three jointly, that is, the ministers, seniors and deacons, is the whole regiment of the church to be committed. . . . Not that we mean to take away the authority of the civil magistrate and chief governor, to whom we wish all blessedness and for the increase of whose godliness we daily pray; but that, Christ being restored into His kingdom, to rule in the same by the scepter of His word and severe discipline, the prince may be better obeyed. . . .

Amend therefore these horrible abuses and reform God's church, and the Lord is on your right hand. . . . Is a reformation good for France, and can it be evil for England? Is discipline meet for Scotland, and is it unprofitable for this realm? Surely God hath set these examples before your eyes to encourage you to go forward to a thorough and a speedy reformation. . . .

141. Separation of Church and State

Some men went further than the Presbyterians, and opposed a state church. Since they advocated separation of church and state, they came to be known as separatists. They were also known as Brownists, after Robert Browne (1550?–1633?), who maintained that each congregation should determine its own ecclesiastical organization and religious practice.

Yet may they [the magistrates] do nothing concerning the church, but only civilly and as civil magistrates: that is, they have not that authority over the church as to be prophets or priests or spiritual kings, as they are magistrates over the same, but only to rule the commonwealth in all outward justice, to maintain the right, welfare and honor thereof with outward power, bodily punishment and civil forcing of men. And therefore also, because the church is in a commonwealth, it is of their charge: that is, concerning the outward provision and outward justice, they are to look to it; but to compel religion, to plant churches by power, and to force a submission to ecclesiastical government by laws and penalties belongeth not to them . . .

142. The Act against Sectaries, 1593

Though Parliament tended to favor the Puritans during the earlier years of Elizabeth's reign, the increasingly radical nature of the move-

---

SOURCES: Robert Browne, "A Treatise of Reformation without Tarrying for Any," *The Writings of Robert Harrison and Robert Browne*, Albert Peel and Leland C. Carlson (eds.), (London: George Allen and Unwin, 1953), 164.

Statutes of the Realm (London: Her Majesty's Stationery Office, 1810–28), IV, 841.

ment alienated moderate men, and in 1593 it passed an act against those who attended religious assemblies other than those in the parish churches. As a result, some went into exile, but, in general, the more extreme manifestations of Puritanism seem to have temporarily died down.

For the preventing and avoiding of such great inconveniences and perils as might happen and grow by the wicked and dangerous practices of seditious sectaries and disloyal persons, be it enacted . . . that, if any person or persons above the age of sixteen years, which shall obstinately refuse to repair to some church, chapel or usual place of common prayer to hear divine service established by her Majesty's laws and statutes in that behalf made, and shall forbear to do the same by the space of a month next after without lawful cause, shall . . . by printing, writing or express words or speeches advisedly and purposely practice or go about to move or persuade any of her Majesty's subjects or any other within her Highness's realms or dominions to deny, withstand and impugn her Majesty's power and authority in causes ecclesiastical . . . ; or to that end or purpose shall advisedly and maliciously move or persuade any other person whatsoever to forbear or abstain from coming to church to hear divine service or to receive the communion according to her Majesty's laws and statutes aforesaid, or to come to or to be present at any unlawful assemblies, conventicles or meetings under color or pretence of any exercise of religion contrary to her Majesty's said laws and statutes; or if any person or persons which shall obstinately refuse to repair . . . to some usual place of common prayer shall . . . willingly join or be present at any such assemblies, conventicles or meetings under color or pretences of any such exercise of religion . . . ; that then every such person so offending as aforesaid and being thereof lawfully convicted shall be committed to prison, there to remain without bail or mainprise[73] until they shall conform and yield themselves to come to some church, chapel or usual place of common prayer and hear divine service according to her Majesty's laws and statutes aforesaid, and to make such open submission and declaration of their said conformity as hereafter in this Act is declared and appointed. . . .

[73] Suretyship for appearance in court.

SECTION J

Defense and Offense

143. The Threat of Mary Stuart, 1586

Beginning in 1569, Mary Stuart, Queen of Scots, was the center of a number of plots to dethrone Elizabeth. In 1586 Anthony Babington was detected in such a conspiracy, which involved the assassination of Elizabeth and an invasion by Philip of Spain in the interest of a Catholic succession. Mary Stuart was involved, and this led to her trial and execution. The following incriminating letter, written by Mary to Babington in July, 1586, reveals her awareness of the worsening Catholic position and of the importance of foreign aid.

Trusty and well-beloved: According to the zeal and entire affection which I have known in you in the common cause of the religion and of my own in particular, I have ever based my hope upon you as a chief and most worthy instrument to be employed in both causes. . . . I cannot but praise, for divers great and important reasons, too long to recite here, your desire to hinder in time the plans of our enemies who seek to destroy our religion in this realm, and ruin all of us together. For long ago I pointed out to the other foreign Catholic princes, and experience has proved me right, that the longer we delayed intervening from both sides the greater advantage we give to our opponents to prevail against the said princes, as they have done against the King of Spain; and meanwhile the Catholics here, exposed to all kinds of persecution and cruelty, steadily grow less in numbers, power and means. . . .

Everything being prepared, and the forces, as well within as without, being ready, then you must set the six gentlemen to work and give order that their design being accomplished, I may be in some way got away from here and that all your forces shall be simultaneously in the field to receive me while we await foreign assistance, which must then be brought up with all speed. Now as no certain day can be appointed for the performance of the said gentlemen's enterprise, I desire there to be always near them, or at least at court, four brave men well horsed to advertize speedily the success of their design, as soon as it is done, to those

---

SOURCE: Alexandre Labanoff, *Lettres, Instructions et Memoires de Marie Stuart* (London: Charles Dolman, 1844), *VI*, 385–96, in J. S. Millward, *Portraits and Documents: Sixteenth Century* (London: Hutchinson Educational, 1961), 96–97.

appointed to get me away from hence, so as to be able to get here before my keeper[74] is informed of the said execution. . . .

This plan seems to me the most suitable for this enterprise, so as to carry it out with care for our own safety. To move on this side before we are sure of good foreign help would simply be to risk to no purpose falling into the same miserable fortune as others who have formerly undertaken in this way. . . .

144. English Involvement in the Netherlands, 1572

Spanish involvement in plots against Elizabeth was one of the factors which led to the deterioration of Anglo-Spanish relations, and finally to out-and-out war. As Spain fished in the troubled waters of English politics, so England intervened to support the Dutch, who had rebelled against their Spanish overlords in 1572. At first English aid was unofficial and voluntary, as is shown in this Spanish communication.

This courier has been delayed and I now write to say that the passion shown by our rebels here is quite incredible. They are with all solicitude sending munitions and money to Flushing and Brille, besides many troops, and they even persuade large numbers of Englishmen to go. The boats after carrying them over return for others, and ship gunpowder, arms, beer and other stores in great quantities, as well as taking over from here and elsewhere grain and other provisions.

A rich English merchant named Pointz, well known in Antwerp, has gone to Flushing with all the money he could collect from the heretic congregations, as well as quantities of arms and munitions. He writes every day to the rebels here to send more help, which they do.

An Englishman named Captain Morgan has enlisted three hundred English soldiers (although without drum and standard), and is ready to leave with them, the citizens of London guaranteeing them their pay, which the Flemish rebels here have undertaken to provide. It is said that many more Englishmen will go, and all this is done so publicly that one is bound to believe that the Queen and Council willingly shut their eyes to it. No doubt great aid will be sent from here daily. God grant, at least, that such aid may not be publicly declared by the state. . . .

Source: Martin A. S. Hume (ed.), *Calendar of Letters and State Papers, Spanish* (London: Her Majesty's Stationery Office, 1892–99), *II*, 391–92.

---

[74] Sir Amias Paulet, later a commissioner on Mary's trial.

145. The Treaty with the Dutch Provinces, 1585

In 1585 the Elizabethan government, alarmed by Spanish successes in the Netherlands, committed itself to a formal alliance with the United (Dutch) Provinces. Careful stipulations were made for reimbursal for its outlays.

1. That the Queen of England should send to the United Provinces an aid of . . . 5000 footmen and one thousand horse, under the conduct of a Governor-General who should be a person of quality and rank, well affected to the true religion, and under other good chiefs and captains; all of whom should be paid by the Queen, so long as the war lasts.

2. . . . The United Provinces, individually and collectively, bind themselves when, by God's grace and her Majesty's assistance, they shall be re-established in peace and repose, to repay all that her Majesty shall have disbursed, as well for the levy of the troops and their transportation, as for their wages. . . .

3. For greater assurance of the repayment, the town of Flushing, the castle of Rammekens in the isle of Walcheren, and the town of Brille, with two fortresses in Holland, shall within one month of the confirmation of the contract be placed in the hands of such governors as it shall please her Majesty to appoint, to be kept by garrisons of her troops until her Majesty shall be completely repaid. . . .

11. The commanders and governors of her Majesty's garrisons shall take an oath of loyalty both to her Majesty and to the States General. . . .

16. Her Majesty may introduce two of her subjects, besides the Governor who will for his part be there, into the Council of State. . . .

21. The States, as well in general as severally, shall not treat with the enemy without her Majesty's knowledge and consent. . . .

22. Her Majesty will also be pleased not to treat of peace or cause to be treated, with the King of Spain or other enemies of the States in any matter touching the United Provinces . . . without the advice and assent of the States General. . . .

25. Whensoever her Majesty for the common defense shall send ships of war to sea against any enemy fleet that may enter the strait between France and England and the United Provinces, the States shall then also equip an equal number of ships of war. . . .

SOURCE: Jean Dumont, *Corps Universel Diplomatique du Droit des Gens* (Amsterdam: P. Brunel [etc.], 1726–31), V, 454–55, in R. B. Wernham and J. C. Walker, *England under Elizabeth, 1558–1603* (London: Longmans, Green, 1932), 58–59.

146. The Armada Crisis, 1588

Spain's attempt in the summer of 1588 to gain control of the English Channel, so that Spanish troops in the Netherlands, as well as others brought in her fleet, might invade England, posed a mighty threat. The rout of the Armada is described for us by Robert Carey, a young courtier involved in the naval action, who notes the part played in the final outcome by the lack of munitions and the storm.

The next year [1588] the King of Spain's great Armada came upon our coast, thinking to devour us all. Upon the news sent to court from Plymouth of their certain arrival, my Lord Cumberland[75] and myself took post horse and rode straight to Portsmouth, where we found a frigate that carried us to sea; and having sought for the fleets a whole day, the night after we fell amongst them, where it was our fortune to light first on the Spanish fleet; and finding ourselves in the wrong, we tacked about and in some short time got to our own fleet, which was not far from the other. . . .

It was on Thursday that we came to the fleet. All that day we followed close the Spanish Armada, and nothing was attempted on either side; the same course we held all Friday and Saturday, by which time the Spanish fleet cast anchor just before Calais. We likewise did the same, a very small distance behind them, and so continued till Monday morning about two of the clock; in which time our council of war had provided six old hulks, and stuffed them full of all combustible matter fit for burning, and on Monday at two in the morning they were let loose, with each of them a man in her to direct them. The tide serving, they brought them very near the Spanish fleet, so that they could not miss to come amongst the midst of them; then they set fire on them and came off themselves, having each of them a little boat to bring him off. The ships set on fire came so directly to the Spanish fleet as they had no way to avoid them but to cut all their hawsers and so escape; and their haste was such that they left one of their four great galleasses on ground before Calais, which our men took and had the spoil of, where many of the Spaniards were slain with the Governor thereof, but most of them were saved with wading ashore to Calais. They being in this disorder, we made ready to follow them, where began a cruel fight, and we had such advantage both of wind and tide as we had a glorious day of them, continuing fight from four o'clock

---

SOURCE: *Memoirs of Robert Cary, Earl of Monmouth, Written by Himself* (Edinburgh: Archibald Constable, 1808; London: John Murray), 15–19.

---

75 George Clifford, Earl of Cumberland, who commanded one of the Queen's ships against the Armada.

in the morning till almost five or six at night, where they lost a dozen or fourteen of their best ships, some sunk, and the rest ran ashore in divers parts to keep themselves from sinking.

After God had given us this great victory they made all the haste they could away and we followed them Tuesday and Wednesday, by which time they were gotten as far as Flamborough Head. It was resolved on Wednesday at night that by four o'clock on Thursday we should have a new fight with them for a farewell; but by two in the morning there was a flag of council hung out in our Vice-Admiral, when it was found that in the whole fleet there was not munition sufficient to make half a fight; and therefore it was there concluded that we should let them pass, and our fleet to return to the Downs. That night we parted with them we had a mighty storm. Our fleet cast anchor and endured it, but the Spanish fleet, wanting their anchors, were many of them cast ashore on the west of Ireland, where they had all their throats cut by the Kernes,[76] and some of them on Scotland, where they were no better used; and the rest (with much ado) got into Spain again. Thus did God bless us and gave victory over this invincible navy; the sea calmed and all our ships came to the Downs on Friday in safety. . . .

SECTION K

The Dawn of Imperial Expansion

147. The Promotion of American Colonies, 1584

Englishmen began to become aware of the advantages of colonies or, to use their term, *plantations,* in the time of Elizabeth. Particularly notable as an advocate of overseas settlement was Richard Hakluyt, best known for his *Principal Navigations, Voyages, and Discoveries of the English Nation* (1589). Five years earlier Hakluyt wrote a treatise, entitled *Discourse of Western Planting,* in an attempt to induce Queen Elizabeth to promote colonization in America. The following chapter titles from that work suggest the major arguments employed by Hakluyt.

SOURCE: E. G. R. Taylor (ed.), *The Original Writings and Correspondence of the Two Richard Hakluyts* (London: Hakluyt Society, 1935), *II,* 211–13.

---

[76] A term for the poorer class of Irishmen, from whom light-armed foot-soldiers were drawn.

1. That this western discovery will be greatly for the enlargement of the Gospel of Christ, whereunto the princes of the reformed religion are chiefly bound, amongst whom her Majesty is principal.

2. That all other English trades are grown beggarly or dangerous, especially in the King of Spain his dominions, where our men are driven to fling their Bibles and Prayer Books into the sea, and to forswear and renounce their religion and conscience, and consequently their obedience to her Majesty.

3. That this western voyage will yield unto us all the commodities of Europe, Africa and Asia, as far as we were wont to travel, and supply the wants of all our decayed trades.

4. That this enterprise will be for the manifold employment of numbers of idle men, and for breeding[77] of many sufficient, and for utterance[78] of the great quantity of the commodities of our realm.

5. That this voyage will be a great bridle to the Indies of the King of Spain, and a means that we may arrest at our pleasure for the space of ten weeks or three months every year, one or two hundred sail of his subjects' ship at the fishing in Newfoundland.

7. What special means may bring King Philip from his high throne, and make him equal to the princes his neighbors, wherewithal is showed his weakness in the West Indies.

12. That the passage in this voyage is easy and short; that it cutteth not near the trade of any other mighty princes, nor near their countries; that it is to be performed at all times of the year, and needeth but one kind of wind; that Ireland, being full of good havens on the south and west sides, is the nearest part of Europe to it, which by this trade shall be in more security and the sooner drawn to more civility.

13. That hereby the revenues and customs of her Majesty, both outwards and inwards, shall mightily be enlarged by the toll, excises and other duties which without oppression may be raised.

14. That this action will be greatly for the increase, maintenance and safety of our navy, and especially of great shipping, which is the strength of our realm, and for the supportation of those occupations that depend upon the same.

15. That speedy planting in divers fit places is most necessary upon these lucky western discoveries, for fear of the danger of being prevented by other nations which have the like intentions . . .

17. That by these colonies the northwest passage to Cathay[79] and China may easily, quickly and perfectly be searched out, as well by river and overland as by sea . . .

77 Upbringing.
78 Trade.
79 A name originating with Marco Polo, and supposed to refer to northern China.

148. Raleigh's Title to Virginia

Hakluyt's *Discourse* had been written at the request of Sir Walter Raleigh, who played a notable role in American colonial projects. In 1584 he obtained a patent from the Queen authorizing him to take possession of unknown territories in America, in her name.

Elizabeth, by the grace of God. . . . etc. . . . Know ye that of our special grace . . . we have given and granted . . . to our trusty and well-beloved servant, Walter Raleigh, Esquire, and to his heirs and assigns forever, free liberty and license from time to time, and at all times forever hereafter, to discover, search, find out and view such remote, heathen and barbarous lands, countries and territories, not actually possessed of any Christian prince nor inhabited by Christian people, as to him, his heirs and assigns . . . shall seem good, and the same to have, hold, occupy and enjoy . . . forever, with all prerogatives, commodities, jurisdictions, royalties, privileges, franchises and pre-eminences thereto or thereabouts, both by sea and land, whatsoever we by our letters patents may grant . . . or . . . have heretofore granted to any person or persons, bodies politic or corporate . . .

And forasmuch as upon the finding out, discovering or inhabiting of such remote lands, countries and territories as aforesaid, it shall be necessary for the safety of all men that shall adventure themselves in those journeys or voyages to determine to live together in Christian peace and civil quietness, each with other, whereby everyone may with more pleasure and profit enjoy that whereunto they shall attain with great pain and peril, we . . . by these presents do give and grant to the said Walter Raleigh, his heirs and assigns forever, that he and they . . . shall and may from time to time forever hereafter, within the said mentioned remote lands and countries, in the way by the seas thither, and from thence, have full . . . power and authority to correct, punish, pardon, govern and rule by their . . . good discretions and policies, as well in causes capital or criminal, as civil, both marine and other, all such our subjects as shall from time to time adventure themselves in the said journeys or voyages, or that shall at any time hereafter inhabit any such lands . . . as aforesaid, or that shall abide within 200 leagues of any of the said . . . places where the said Walter Raleigh, his heirs or assigns, . . . or any of his or their associates or companies shall inhabit within six years next ensuing the date hereof, according to such statutes, laws and ordinances as shall be

---

SOURCE: Edmund Goldsmid (ed.), *The Principal Navigations . . . of the English Nation, Collected by Richard Hakluyt* (Edinburgh: Goldsmid, 1885–90), *XIII* 276–77, 279–80.

by him the said Walter Raleigh, his heirs and assigns, . . . or any of them devised or established for the better government of the said people as aforesaid. So always as the said statutes, laws and ordinances may be, as near as conveniently may be, agreeable to the form of the laws, statutes, government or policy of England, and also so as they be not against the true Christian faith now professed in the Church of England, nor in any wise to withdraw any of the subjects or people of those lands or places from the allegiance of us, our heirs and successors, as their immediate sovereign under God. . . .

149. Tyrone's Demands, 1599

Englishmen had some success, under Mary and Elizabeth, in establishing plantations in Ireland. But the friction between the native Irish and the English settlers, backed by their government, resulted in much strife and even war. The most serious resistance to English penetration was led by the Irish Earl of Tyrone, and Tyrone's Rebellion (1595–1601) was the gravest problem that Elizabeth had to face during the last years of her reign. The following are extracts from the terms sought by Tyrone in 1599, but regarded by the English government as unacceptable.

1. That the Catholic, Apostolic, and Roman religion be openly preached and taught throughout all Ireland, as well in cities as borough towns, by bishops, seminary priests, Jesuits, and all other religious men.
2. That the Church of Ireland be wholly governed by the Pope.
3. That all cathedrals and parish churches, abbeys, and all other religious houses, with all tithes and church lands, now in the hands of the English, be presently restored to the Catholic churchmen.
4. That all Irish priests and religious men, now prisoners in England or Ireland, be presently set at liberty, with all temporal Irishmen, that are troubled for their conscience, and to go where they will without further trouble.
5. That all Irish priests and religious men may freely pass and repass, by sea and land, to and from foreign countries.
6. That no Englishman may be a churchman in Ireland.
7. That there be erected an university upon the crown rents of Ireland, wherein all sciences shall be taught according to the manner of the Catholic Roman Church.
8. That the Governor of Ireland be at least an Earl, and of the Privy Council of England, bearing the name of Viceroy.

Source: Ernest G. Atkinson (ed.), *Calendar of State Papers Relating to Ireland, 1599–1600* (London: Her Majesty's Stationery Office, 1899), 279–80.

9. That the Lord Chancellor, Lord Treasurer, Lord Admiral, the Council of State, the Justices of the laws, Queen's Attorney, Queen's Serjeant, and all other officers appertaining to the Council and law of Ireland, be Irishmen.

10. That all principal governments of Ireland, as Connaught, Munster, etc., be governed by Irish noblemen.

11. That the Master of Ordnance, and half the soldiers with their officers resident in Ireland, be Irishmen.

12. That no Irishman's heirs shall lose their lands for the faults of their ancestors.

13. That no Irishman's heir under age shall fall in the Queen's or her successors' hands, as a ward,[80] but that the living be put to the heir's profit, and the advancement of his younger brethren, and marriages of his sisters, if he have any.

14. That no children nor any other friends be taken as pledges for the good abearing[81] of their parents, and, if there be any such pledges now in the hands of the English, they must presently be released.

15. That all statutes made against the preferment of Irishmen, as well in their own country as abroad, be presently recalled.

16. That the Queen nor her successors may in no sort press an Irishman to serve them against his will.

17. That O'Neill, O'Donnell, the Earl of Desmond, with all their partakers,[82] may peaceably enjoy all lands and privileges that did appertain to their predecessors 200 years past.

18. That all Irishmen, of what quality they be, may freely travel in foreign countries for their better experience, without making any of the Queen's officers acquainted withal.

19. That all Irishmen may as freely travel and traffic all merchandises in England as Englishmen, paying the same rights and tributes as the English do.

20. That all Irishmen may freely traffic with all merchandises that shall be thought necessary by the Council of State of Ireland for the profit of their republic, with foreigners or in foreign countries, and that no Irishman shall be troubled for the passage of priests or other religious men.

21. That all Irishmen that will may learn and use all occupations and arts whatsoever.

22. That all Irishmen may freely build ships of what burden they will, furnishing the same with artillery and all munition at their pleasure.

80 The crown still exercised the feudal right of wardship, by which it administered the property of certain minors.

81 Behavior.

82 Associates. The reference is to those who had been engaged against the English.

150. The East India Company's First Charter, 1600

English commercial contacts with the Orient were the major objective in most voyages of exploration in the sixteenth century. In 1600 the East India Company obtained its original charter, giving it monopolistic trading rights to the Spice Islands. Its operations were soon diverted to the Indian mainland, and by 1612 the English had gained a foothold at Surat.

Elizabeth, by the grace of God [etc.] . . . Whereas our most dear and loving cousin, George, Earl of Cumberland, and our well-beloved subjects [here over two hundred are named] . . . have of our certain knowledge been petitioners unto us for our royal assent and licence to be granted unto them, that they, at their own adventures, costs and charges, as well for the honor of this our realm of England as for the increase of our navigation and advancement of trade of merchandise within our said realm and the dominions of the same, might adventure and set forth one or more voyages, with convenient number of ships and pinnaces, by way of traffic and merchandise to the East Indies, in the countries and parts of Asia and Africa, and to as many of the islands, ports, cities, towns and places thereabouts, as where trade and traffic of merchandise may by all likelihood be . . . had: . . .

Know ye therefore that we . . . have . . . granted . . . unto our said loving subjects . . . that they . . . from henceforth be . . . one body corporate and politic, in deed and in name, by the name of the Governor and Company of Merchants of London trading into the East Indies, . . . and that by the same name . . . they shall have succession, and that they and their successors . . . shall be at all times hereafter . . . capable in law to have . . . and retain lands, rents, privileges, liberties, jurisdictions, franchises and hereditaments of whatsoever kind . . . ; and also to give . . . and dispose lands, tenements and hereditaments, and to do and execute all and singular other things . . . that to them shall or may appertain to do. . . .

And further we . . . grant unto the said Governor [etc.] . . . that they . . . and all the sons of them . . . at their several ages of one-and-twenty years or upwards, and further all such the apprentices, factors and servants of them . . . which hereafter shall be employed by the said Governor [etc.] . . . may, by the space of fifteen years from the feast of the birth of our Lord God last past before the date hereof, freely traffic and use the trade of merchandise by seas, in and by such ways and passages . . . as

---

SOURCE: John Shaw (ed.), *Charters Relating to the East India Company from 1600 to 1761* (Madras: Government Press, 1887), 1–13.

they shall esteem and take to be fittest, into and from the said East Indies, in the countries and parts of Asia and Africa, and into and from all the islands . . . and places of Asia and Africa and America, . . . beyond the Cape of Bona Esperanza[83] to the Straits of Magellan, . . . in such order . . . as shall be, from time to time, . . . limited and agreed, . . . any statute, usage, diversity of religion or faith, or any other cause . . . to the contrary notwithstanding: so always the same trade be not undertaken nor addressed to any country . . . or place already in the lawful and actual possession of any such Christian prince or state as at this present is or at any time hereafter shall be in league or amity with us, our heirs and successors, and who doth not or will not accept of such trade but doth overtly declare and publish the same to be utterly against his or their good-will and liking. . . .

And we . . . grant to the said Governor [etc.] . . . that the said East Indies . . . shall not be visited . . . by any of the subjects of us, our heirs or successors during the same term of fifteen years, contrary to the true meaning of these presents; and by virtue of our prerogative royal, which we will not in that behalf have argued or brought in question, we straitly charge . . . all the subjects of us, our heirs and successors, . . . that none of them, directly or indirectly, do visit . . . or trade . . . into or from any of the said East Indies or into or from any . . . the places aforesaid, other than the said Governor [etc.] . . . unless it be by and with such license and agreement of the said Governor [etc.] . . . upon pain that every such person . . . shall incur our indignation and the forfeiture and loss of the goods . . . which so shall be brought into this realm . . . contrary to our said prohibition, . . . as also the . . . ships, with the furniture thereof, wherein such goods . . . shall be brought . . . ; and further, all . . . the said offenders . . . to suffer imprisonment during our pleasure, and such other punishment as to us . . . shall seem meet and convenient . . .

And the said Governor [etc.] . . . do . . . promise and grant to and with us . . . that they . . . in all and every such voyages as they . . . shall make out of this realm by virtue of this our grant . . . (the first voyage only excepted), . . . will, upon every return which shall be made back again into this realm, or any of our dominions, or within six months next after every such return, bring into this our realm . . . as great or greater value in bullion of gold or silver, or other foreign coin of gold or silver, . . . as shall be by force of these presents transported and carried out of this realm. . . .

[83] Cape of Good Hope.

SECTION L

Socio-economic Problems and Policies

151. Sir Thomas More on Enclosures

Under the Tudors, and particularly during the earlier part of the era, the enclosure of land caused widespread complaint. Land enclosure, or the agglomeration of small holdings in larger estates, was a manifestation of the increased capitalistic organization of agriculture. The dispossession of small tenants, and, where enclosure was connected with sheep-raising, the reduction of the rural labor force were inevitable results. Among the many indictments of the movement, we may select passages from More's *Utopia.*

Forsooth, . . . your sheep, that were wont to be so meek and tame and so small eaters, now, as I hear say, be become so great devourers and so wild that they eat up and swallow down the very men themselves. They consume, destroy and devour whole fields, houses and cities. For look in what parts of the realm doth grow the finest and therefore dearest wool, there noblemen and gentlemen – yea, and certain abbots, holy men, no doubt – not contenting themselves with the yearly revenues and profits that were wont to grow to their forefathers and predecessors of their lands, nor being content that they live in rest and pleasure, nothing profiting, yea, much noying[84] the weal public, leave no ground for tillage; they enclose all in pastures; they throw down houses; they pluck down towns, and leave nothing standing but only the church, to make of it a sheephouse. And, as though you lost no small quantity of ground by forests, chases, lawns and parks, those good holy men turn all dwelling places and all glebe land into desolation and wilderness.

Therefore, that one covetous and insatiable cormorant and very plague of his native country may compass about and enclose many thousand acres of ground together within one pale or hedge, the husbandmen be thrust out of their own; or else either by covin[85] or fraud, or by violent oppression, they be put besides it,[86] or by wrongs and injuries they be

SOURCE: J. H. Lupton (ed.), *The Utopia of Sir Thomas More* (Oxford: The Clarendon Press, 1895), 51–54.

[84] Troubling.
[85] Deceit.
[86] Ousted.

so wearied that they be compelled to sell all. By one means therefore or by other, either by hook or crook, they must needs depart away, poor, silly, wretched souls – men, women, husbands, wives, fatherless children, widows, woeful mothers with their young babes, and their whole household, small in substance and much in number, as husbandry requireth many hands. Away they trudge, I say, out of their known and accustomed houses, finding no place to rest in. All their household stuff, which is very little worth . . . they be constrained to sell it for a thing of nought. And when they have, wandering about, soon spent that, what can they else do but steal, and then . . . be hanged, or else go about a-begging? And yet then also they be cast in prison as vagabonds, because they go about and work not. . . .

152. Ket's Rebellion, 1549

In 1549 there was a peasant rising in Norfolk against enclosures, known as Ket's Rebellion. Led by Robert Ket, himself a landed proprietor, some sixteen thousand peasants destroyed hedges, plundered landowners, and took some of them prisoners, before being overcome by the use of armed force. Their demands were set forth, in part, as follows.

We certify your Grace that whereas the lords of the manors hath been charged with certe free rent,[87] the same lords hath sought means to charge the freeholders to pay the same rent, contrary to right.

We pray your Grace that no lord of no manor shall common upon the commons.[88]. . .

We pray that reed ground and meadow ground may be at such price as they were in the first year of King Henry VII.

We pray that all marshes that are holden of the King's Majesty by free rent or of any other, may be again at the price that they were in the first year of King Henry VII. . . .

We pray that all freeholders and copyholders may take the profits of all commons, and there to common, and the lords not to common nor take profits of the same. . . .

We pray that copyhold land that is unreasonable rented may go as it

SOURCE: Frederick W. Russell, *Kett's Rebellion in Norfolk* (London: Longmans, Green, 1859), 48–56.

[87] The rent due from the lord of a manor to a superior lord under whom he held it, which he was bound to pay himself and not exact from others.
[88] Have rights in the common land.

did in the first year of King Henry VII, and that at the death of a tenant or at a sale the same lands to be charged with an easy fine, as a capon or a reasonable [sum] of money for a remembrance. . . .

We pray your Grace to give license and authority by your gracious commission under your great seal to such commissioners as your poor commons hath chosen, or to as many of them as your Majesty and your council shall appoint and think meet, for to redress and reform all such good laws, statutes, proclamations and all other your proceedings, which hath been hidden by your justices of your peace, sheriffs, escheators[89] and other your officers from your poor commons, since the first year of the reign of your noble grandfather King Henry VII.

We pray that those your officers that hath offended your Grace and your commons, and so proved by the complaint of your poor commons, do give unto those poor men so assembled 4d every day so long as they have remained there.

We pray that no lord, knight, esquire nor gentleman do graze nor feed any bullocks or sheep if he may spend forty pounds a year by his lands, but only for the provision of his house.

By me, *Robert Ket*
By me, *Thomas Aldryche* *Thomas Cod*

153. The Statute of Artificers, 1563

The government of Queen Elizabeth concerned itself to a notable extent with the regulation of social and economic conditions. In 1563 Parliament passed the Statute of Artificers. This was a kind of labor code, designed to contribute stability to a situation which had become somewhat chaotic because of enclosures, displacement of population, and the lessening effectiveness of the craft guilds. It was by far the most ambitious statement of Tudor economic regulation.

Although there remain and stand in force presently a great number of acts and statutes concerning the retaining, departing, wages and orders of apprentices, servants and laborers, as well in husbandry as in divers other arts, mysteries and occupations, yet, partly for the imperfection and contrariety that is found and do appear in sundry of the said laws, and for the variety and number of them, and chiefly for that the wages and allowances limited and rated in many of the said statutes are in divers

---

Source: *Statutes of the Realm* (London: Her Majesty's Stationery Office, 1810–28), *IV*, 414–21.

---

89 Officials called upon to attend to the escheats (reverting lands) of the crown.

places too small and not answerable to this time, respecting the advancement of prices of all things belonging to the said servants and laborers, the said laws cannot conveniently, without the great grief and burden of the poor laborer and hired man, be put in good and due execution; and as the said several acts and statutes were at the time of the making of them thought to be very good and beneficial for the commonwealth of this realm, as divers of them yet are, so if the substance of as many of the said laws as are meet to be continued shall be digested and reduced into one sole law and statute, and in the same an uniform order prescribed and limited concerning the wages and other orders for apprentices, servants and laborers, here is good hope that there will come to pass that the same law, being duly executed, should banish idleness, advance husbandry, and yield unto the hired person both in the time of scarcity and in the time of plenty a convenient proportion of wages . . . :

And be it . . . enacted that no person which shall retain any servant shall put away his or her said servant, and that no person retained according to this statute shall depart from his master, mistress or dame before the end of his or her term . . . , unless it be for some reasonable and sufficient cause or matter to be allowed before two justices of peace, or one at the least, within the . . . county, or before the mayor or other chief officer of the city, borough or town corporate wherein the said master, mistress or dame inhabiteth, to whom any of the parties grieved shall complain . . .

And be it further enacted . . . that every person between the age of twelve years . . . and three score years, not [otherwise lawfully employed] . . . shall after the . . . last day of September now next ensuing . . . be compelled to be retained to serve in husbandry by the year . . .

And for the declaration and limitation what wages servants, laborers and artificers, either by the year or day or otherwise, shall have and receive, be it enacted . . . that the justices of peace . . . shall yearly, at every general sessions first to be holden and kept after Easter . . . , assemble themselves together; and they so assembled, calling unto them such discreet and grave persons . . . as they shall think meet, and conferring together respecting the plenty or scarcity of the time and other circumstances necessary to be considered, shall have authority . . . to limit, rate and appoint the wages . . . of . . . artificers, handicraftsmen, husbandmen, or any other laborer, servant or workman . . . , and shall . . . certify the same, engrossed in parchment with the considerations and causes thereof under their hands and seals, into the Queen's most honorable Court of Chancery . . .

Provided always . . . that in the time of hay or corn harvest, the justices of peace . . . and also the constable or other head officer of every township, upon request and for the avoiding of the loss of any corn, grain or hay, shall and may cause all such artificers and persons as be meet to labor . . . to serve by the day for the mowing, reaping, shearing, getting or inning[90] of corn, grain and hay . . .

And be it further enacted . . . that two justices of peace, the mayor . . . of any city, borough or town corporate, and two aldermen or two other discreet burgesses . . . shall and may . . . appoint any such woman as is of the age of twelve years and under the age of forty years and unmarried and forth[91] of service, as they shall think meet to serve, to be retained or serve by the year or by the week or day, for such wages and in such reasonable sort and manner as they shall think meet. . . .

And if any . . . master shall misuse or evil intreat his apprentice, or . . . the said apprentice shall have any just cause to complain, or the apprentice do not his duty to his master, then the said master or prentice being grieved and having cause to complain shall repair unto one justice of peace within the said county, or to the mayor . . . of the city, town corporate, market town or other place where the said master dwelleth, who shall by his wisdom and discretion take such order and direction between the said master and his apprentice as the equity of the cause shall require. . . .

154. The Poor Law of 1598

Prior to the sixteenth century the relief of poverty was regarded as the province of the church or of private individuals, but with the destruction of the monasteries and confiscation of other church property the state was forced to assume some responsibility. Beginning with Henry VIII, Parliament enacted legislation on the subject, which may be said to be summed up in the Poor Law of 1598, a measure reenacted in 1601 and not formally repealed until the nineteenth century. The principal objectives of the act are found in this introductory section.

Be it enacted by the authority of this present Parliament that the churchwardens of every parish and four substantial householders there . . . , who shall be nominated yearly in Easter week under the hand and seal of two or more justices of the peace in the same county . . . dwelling in or near the same parish, shall be called overseers of the poor of the same parish; and they, or the greater part of them, shall take order from time

---

SOURCE: *Statutes of the Realm* (London: Her Majesty's Stationery Office, 1810–28), IV, 896.

---

90 Taking in.
91 Out of.

to time, by and with the consent of two or more such justices of the peace, for setting to work of the children of all such whose parents shall not by the said persons be thought able to keep and maintain their children; and also all such persons married and unmarried as, having no means to maintain them, use no ordinary and daily trade of life to get their living by; and also to raise weekly or otherwise, by taxation of every inhabitant and every occupier of lands in the said parish in such competent sum and sums of money as they shall think fit, a convenient stock of flax, hemp, wool, thread, iron and other necessary ware and stuff to set the poor on work, and also competent sums of money for and towards the necessary relief of the lame, impotent, old, blind and such other among them being poor and not able to work, and also for the putting out of such children to be apprentices, to be gathered out of the same parish according to the ability of the said parish; and to do and execute all other things, as well for disposing of the said stock as otherwise concerning the premises, as to them shall seem convenient. . . .

155. Opposition to Monopolies

In the latter part of Elizabeth's reign sharp protests were voiced against the royal practice of granting monopolies of the sale of established products to certain individuals, either as rewards or for a consideration. This tended to raise the price paid by the consumer. The matter was debated in Parliament, and the Queen, sensing popular resentment, was prevailed upon to discontinue some of the existing patents. Some passages of the Commons debate of 1601 follow.

Sir Edward Stanhope informed the House of the great abuse by the patentee of salt in his country . . . "To Lynn there is every year brought at least 3000 weight of salt, and every weight, since this patent, is enhanced twenty shillings; and where the bushel was wont to be eight pence, it is now sixteen pence. And, I dare boldly say it, if this patent were called in, there might well be £3000 a year saved in the ports of Lynn, Boston and Hull. I speak this of white salt.". . .

Mr. Solicitor Fleming said, "I will briefly give you an account of all things touching these monopolies. Her Majesty, in her provident care, gave charge to Mr. Attorney and myself, that speedy and special course may be taken for these patents. This was in the beginning of Hilary

---

SOURCE: Hayward Townshend, *Historical Collections, or an Exact Account of the Proceedings of the Four Last Parliaments of Queen Elizabeth* (London: T. Basset [etc.], 1680), 238–39.

Term[92] last. But you all know the danger of that time,[93] and what great affairs of importance happened to prevent that business. Since that, nothing could be done therein for want of leisure."

Sir Robert Wroth said, "I would but note, Mr. Solicitor, that you were charged to take care in Hilary Term last. Why not before? There was time enough ever since the last Parliament. I speak it and I speak it boldly: these patentees are worse than ever they were . . .

"There have been divers patents granted since the last Parliament. These are now in being, viz.: the patents for currants, iron, powder, cards, horns, ox shin-bones, train oil, lists of cloth, ashes, bottles, glasses, bags, shreds[94] of gloves, aniseed, vinegar, sea coals, steel, aqua vitae, brushes, pots, salt, saltpeter, lead, accidence,[95] oil, transportation of leather, calamint stone, oil of blubber, fumothoes (or dried pilchers in the smoke), and divers others."

Upon reading of the patents aforesaid, Mr. Hackwell of Lincoln's Inn stood up and asked this, "Is not bread there?" "Bread?" quoth another . . . "No," quoth Mr. Hackwell, "but if order be not taken for these bread will be there before the next Parliament.". . .

SECTION M

Tudor Government

156. The Authority of Parliament

Although the Tudor monarchy is rightly identified with the strong personal rule of the sovereign, the sixteenth century witnessed a notable growth in the importance of Parliament. The power of the House of Commons came to surpass that of the Lords, and men sought, as they had never done in the past, to be elected to it. The contemporary scholar and statesman, Sir Thomas Smith, has this to say of Parliament in his *De Republica Anglorum,* published in 1589.

SOURCE: Thomas Smith, *De Republica Anglorum, a Discourse on the Commonwealth of England,* L. Alston (ed.), (London: Cambridge University Press, 1906), 48–49.

92 One of the four terms in which the law courts sat. It began January 11, St. Hilary's Day.

93 The reference is to the abortive coup d'etat of the Earl of Essex.

94 Strips used in making gloves.

95 Accidence: a book on rudimentary grammar.

The most high and absolute power of the realm of England consisteth in the Parliament. For as in war where the King himself in person, the nobility, the rest of the gentility, and the yeomanry are, is the force and power of England; so in peace and consultation where the prince is to give life, and the last and highest commandment, the barony for the nobility and higher, the knights, esquires, gentlemen and commons for the lower part of the commonwealth, the bishops for the clergy, be present to advertize, consult and show what is good and necessary for the commonwealth, and to consult together, and upon mature deliberation every bill or law being thrice read and disputed upon in either House, . . . and after the prince himself . . . doth consent unto and alloweth. That is the prince's and the whole realm's deed, whereupon justly no man can complain, but must accommodate himself to find it good and obey it.

That which is done by this consent is called firm, stable and sanctum,[96] and is taken for law. The Parliament abrogateth old laws, maketh new, giveth orders for things past and for things hereafter to be followed, changeth rights and possessions of private men, legitimateth bastards, establisheth forms of religion, altereth weights and measures, giveth forms of succession to the crown, defineth of doubtful rights, whereof is no law already made, appointeth subsidies, tailes,[97] taxes and impositions, giveth most free pardons and absolutions, restoreth in blood and name as the highest court, condemneth or absolveth them whom the prince will put to that trial.

And to be short, all that ever the people of Rome might do, either *in centuriatis comitiis* or *tributis*,[98] the same may be done by the Parliament of England, which representeth and hath the power of the whole realm, both the head and the body. For every Englishman is intended to be there present, either in person or by procuration and attorneys, of what pre-eminence, state, dignity or quality soever he be, from the prince (be he King or Queen) to the lowest person of England. And the consent of the Parliament is taken to be every man's consent.

157. Privileges of the Commons, 1562

As the House of Commons grew in importance, it became increasingly aware of its privileges. During the early years of Queen Elizabeth it

SOURCE: Sir Simonds D'Ewes, *A Complete Journal of the Votes, Speeches, and Debates . . . throughout the Whole Reign of Queen Elizabeth* (London: Paul Bowes, 1693), 65–66.

[96] Established as inviolable.
[97] Taxes.
[98] In the centuriate assembly or the tribal assembly.

became the regular practice for the Speaker to request that the sovereign recognize three great privileges: freedom of access, freedom of debate, and freedom from arrest. Speaker Williams used these words in 1562:

Further, I am to be a suitor to your Majesty that, when matters of importance shall arise whereupon it shall be necessary to have your Highness' opinion, that then I may have free access unto you for the same; and the like to the lords of the upper house.

Secondly, that in repairing from the nether house to your Majesty, or the lords of the upper house, to declare their meanings, and I mistaking or uttering the same contrary to their meaning, that then my fault or imbecility in declaring thereof be not prejudicial to the house, but that I may again repair to them, the better to understand their meanings and so they to reform the same.

Thirdly, that the assembly of the lower house may have frank and free liberties to speak their minds without any controlment, blame, grudge, menaces or displeasure, according to the old ancient order.

Finally, that the old privilege of the house be observed, which is that they and theirs might be at liberty, frank and free, without arrest, molestation, trouble or other damage to their bodies, lands, goods or servants, with all other their liberties, during the time of the said Parliament, whereby they may the better attend and do their duty; all which privileges I desire may be enrolled, as at other times it hath been accustomed. . . .

158. Freedom of Debate, 1576

Despite its claims, the House of Commons did not enjoy unrestricted freedom of speech under Elizabeth. On various occasions she prohibited debate on her marriage, the succession to the throne, and certain ecclesiastical matters. Peter Wentworth, a Member who persisted in discussing forbidden questions, paid for his audacity with sojourns in prison, once upon the order of his colleagues. In a speech in 1576, Wentworth upheld the importance of the privilege in the following words.

Mr. Speaker, I find written in a little volume these words, in effect: sweet is the name of liberty, but the thing itself a value beyond all inestimable treasure. So much the more it behoveth us to take care lest we,

---

SOURCE: Sir Simonds D'Ewes, *A Complete Journal of the Votes, Speeches, and Debates . . . throughout the Whole Reign of Queen Elizabeth* (London: Paul Bowes, 1693), 236–37.

contenting ourselves with the sweetness of the name, lose and forego the thing, being of the greatest value that can come into this noble realm. The inestimable treasure is the use of it in this House. . . .

Sometime it happeneth that a good man will in this place (for argument sake) prefer an evil cause, both for that he would have a doubtful truth to be opened and manifested, and also the evil prevented; so that to this point I conclude, that in this House, which is termed a place of free speech, there is nothing so necessary for the preservation of the prince and state as free speech, and without it is a scorn and mockery to call it a Parliament house, for in truth it is none, but a very school of flattery and dissimulation, and so a fit place to serve the devil and his angels in, and not to glorify God and benefit the commonwealth. . . .

Amongst other, Mr. Speaker, two things do great hurt in this place, of the which I do mean to speak. The one is a rumor which runneth about the House, and this it is: take heed what you do, the Queen's Majesty liketh not such a matter, whosoever prefereth it, she will be offended with him; or the contrary: her Majesty liketh of such a matter, whosoever speaketh against it, she will be much offended with him.

The other: sometimes a message is brought into the House, either of commanding or inhibiting, very injurious to the freedom of speech and consultation. I would to God, Mr. Speaker, that these two were buried in hell, I mean rumors and messages, for wicked undoubtedly they are; the reason is, the devil was the first author of them, from whom proceedeth nothing but wickedness . . .

159. The Queen's Aides

In the conduct of her administration, Elizabeth, as well as other Tudor sovereigns, relied primarily upon the privy councillors and the justices of the peace. The former served as her principal advisers and aides at the center of government, while the latter were her chief agents locally. Below is the oath of a privy councillor, as found in state papers of Elizabeth's time, and Sir Thomas Smith's description of the functions of the justices.

Oath of a Privy Councillor — You shall swear to be a true and faithful councillor to the Queen's Majesty as one of her Highness' Privy Council.

SOURCES: G. W. Prothero (ed.), *Select Statutes and Other Constitutional Documents Illustrative of the Reigns of Elizabeth and James I* (Oxford: The Clarendon Press, 1906), 165–66.

Thomas Smith, *De Republica Anglorum, a Discourse on the Commonwealth of England,* L. Alston (ed.), (London: Cambridge University Press, 1906), 86, 88.

You shall not know or understand of any manner [of] thing to be attempted, done or spoken against her Majesty's person, honor, crown or dignity royal, but you shall let and withstand the same to the uttermost of your power, and either do or cause it to be forthwith revealed either to her Majesty's self or to the rest of her Privy Council. You shall keep secret all matters committed and revealed to you as her Majesty's councillor or that shall be treated of secretly in council. And if any of the same treaties or counsels shall touch any other of the councillors, you shall not reveal the same to him, but shall keep the same until such time as by the consent of her Majesty or of the rest of the council publication shall be made thereof. You shall not let[99] to give true, plain and faithful counsel at all times, without respect either of the cause or of the person, laying apart all favor, need, affection and partiality. And you shall to your uttermost bear faith and true allegiance to the Queen's Majesty, her heirs and lawful successors, and shall assist and defend all jurisdictions, pre-eminences and authorities granted to her Majesty and annexed to her crown, against all foreign princes, persons, prelates or potentates, whether by act of parliament or otherwise. And generally in all things you shall do as a faithful and true councillor ought to do to her Majesty. So help you God and the holy contents of this book.

The Functions of the Justice of the Peace – The Justices of the Peace be those in whom at this time, for the repressing of robbers, thieves and vagabonds, of privy complots[100] and conspiracies, of riots and violences and all other misdemeanors in the commonwealth, the prince putteth his special trust. Each of them hath authority upon complaint to him made of any theft, robbery, manslaughter, murder, violence, complots, riots, unlawful games, or any such disturbance of the peace and quiet of the realm, to commit the persons whom he supposeth offenders to the prison, and to charge the constable or sheriff to bring them thither, the gaoler to receive them and keep them till he and his fellows do meet. A few lines signed with his hand is enough for that purpose. These do meet four times in the year, that is, in each quarter once, to inquire of all the misdemeanors aforesaid . . . These meetings . . . be called quarter sessions . . .

The Justices of the Peace do meet also at other times by commandment of the prince upon suspicion of war, to take order for the safety of the shire, sometimes to take musters of harness[101] and able men, and sometimes to take orders for the excessive wages of servants and laborers, for excess of apparel, for unlawful games, for conventicles and evil orders in

[99] Forbear.
[100] Conspiracies.
[101] Military equipment.

alehouses and taverns, for punishment of idle and vagabond persons; and generally, as I have said, for the good government of the shire the prince putteth his confidence in them. . . .

160. Elizabeth's Last Speech to the Commons, 1601

Although Elizabeth's relations with her Parliaments had their ups and downs, and tended to become more difficult in her later years, she retained to the end her skill as a politician and a sort of sixth sense which kept her aware of the popular temper. Few could have doubted her dedication to the welfare of Englishmen, so apparent in these excerpts from her last speech to the Commons.

Mr. Speaker, we have heard your declaration and perceive your care of our state, by falling into the consideration of a grateful acknowledgment of such benefits as you have received; and that your coming is to present thanks unto us, which I accept with no less joy than your loves can have desire to offer such a present.

I do assure you that there is no prince that loveth his subjects better, or whose love can countervail our love; there is no jewel, be it of never so rich a price, which I prefer before this jewel, I mean your love, for I do more esteem it than any treasure or riches; for that we know how to prize, but love and thanks I count inestimable. And though God hath raised me high, yet this I count the glory of my crown, that I have reigned with your loves. This makes me that I do not so much rejoice that God hath made me to be a queen as to be a queen over so thankful a people. Therefore I have cause to wish nothing more than to content the subject, and that is a duty which I owe. Neither do I desire to live longer days than that I may see your prosperity, and that is my only desire. And as I am that person that still, yet under God, hath delivered you, so I trust by the almighty power of God that I still shall be his instrument to preserve you from envy, peril, dishonor, shame, tyranny and oppression, partly by means of your intended helps, which we take very acceptably, because it manifesteth the largeness of your loves and loyalties unto your sovereign.

Of myself I must say this: I never was any greedy, scraping grasper, nor a strait[102] fast-holding prince, nor yet a waster; my heart was never

---

SOURCE: William Cobbett, *Parliamentary History of England* (London: R. Bagshaw [etc.], 1806–20), *I*, 940–42.

---

[102] Strict.

set on worldly goods, but only for my subjects' good. What you do bestow on me, I will not hoard it up but receive it to bestow on you again. Yea, mine own properties I count yours, to be expended for your good. . . .

I know the title of a king is a glorious title; but assure yourself that the shining glory of princely authority hath not so dazzled the eyes of our understanding but that we well know and remember that we also are to yield an account of our actions before the great judge. To be a king and wear a crown is more glorious to them that see it than it is pleasure to them that bear it. For myself, I was never so much enticed with the glorious name of a king, or royal authority of a queen, as delighted that God hath made me His instrument to maintain His truth and glory, and to defend this kingdom (as I said) from peril, dishonor, tyranny and oppression. There will never queen sit in my seat with more zeal to my country, or care to my subjects, and that will sooner, with willingness, yield and venture her life for your good and safety than myself. And though you have had, and may have, many princes more mighty and wise, sitting in this seat, yet you never had, or shall have, any that will be more careful and loving . . .

And so I commit you all to your best fortunes and further counsels. And I pray you, Mr. Comptroller, Mr. Secretary, and you of my council, that before these gentlemen depart into their countries you bring them all to kiss my hand.

Part 5

THE STUART ERA

1603-1689

SECTION A

Political Friction under James I

161. James's Speech on the Royal Power, 1610

James I and his son, Charles I, are more closely identified with the theory of divine right of kings than any other English monarchs. The theory had existed long before their day, but James in particular defined and expounded it as a practical basis of government. The following remarks are taken from one of James's speeches to Parliament.

The state of monarchy is the supremest thing upon earth, for kings are not only God's lieutenants upon earth and sit upon God's throne, but even by God himself they are called gods. There be three principal similitudes that illustrate the state of monarchy: one taken out of the word of God, and the two other out of the grounds of policy and philosophy. In the Scriptures kings are called gods, and so their power after a certain relation compared to the divine power. Kings are also compared to fathers of families, for a king is truly *parens patriae*,[1] the politic father of his people. And lastly, kings are compared to the head of this microcosm of the body of man . . .

I conclude then this point touching the power of kings with this axiom of divinity: that as to dispute what God may do is blasphemy, . . . so it is sedition in subjects to dispute what a king may do in the height of his

SOURCE: *Works of James I* (London: R. Barker and J. Bill, 1616), 529–37, in G. W. Prothero (ed.), *Select Statutes and Other Constitutional Documents Illustrative of the Reigns of Elizabeth and James I* (Oxford: The Clarendon Press, 1906), 293–95.

[1] Father of his country.

power. But just kings will ever be willing to declare what they will do, if they will not incur the curse of God. I will not be content that my power be disputed upon, but I shall ever be willing to make the reason appear of all my doings, and rule my actions according to my laws . . .

Now the second general ground whereof I am to speak concerns the matter of grievances. . . . First then, I am not to find fault that you inform yourselves of the particular just grievances of the people; nay I must tell you, ye can neither be just nor faithful to me or to your countries that trust and employ you, if you do it not . . . But I would wish you to be careful to avoid three things in the matter of grievances.

First, that you do not meddle with the main points of government: that is my craft . . . ; to meddle with that were to lesson me. I am now an old king . . . ; therefore there should not be too many Phormios[2] to teach Hannibal: I must not be taught my office.

Secondly, I would not have you meddle with such ancient rights of mine as I have received from my predecessors . . . : such things I would be sorry should be accounted for grievances. All novelties are dangerous, as well in a politic as in a natural body; and therefore I would be loath to be quarrelled in my ancient rights and possessions, for that were to judge me unworthy of that which my predecessors had and left me.

And lastly, I pray you beware to exhibit for grievance anything that is established by a settled law, and whereunto (as you have already had a proof) you know I will never give a plausible[3] answer; for it is an undutiful part in subjects to press their king, wherein they know beforehand he will refuse them. Now, if any law or statute be not convenient, let it be amended by Parliament, but in the meantime term it not a grievance; for to be grieved with the law is to be grieved with the king, who is sworn to be the patron and maintainer thereof. . . .

162. The Apology of the Commons, 1604

The House of Commons had its own concepts of its rightful functions, which clashed with King James's divine right position. In 1604 the Commons prepared an apology, or defense, of its rights and objectives. Though never presented to James, it is important in revealing the

---

SOURCE: William Petyt, *Jus Parliamentarium* (London: J. Nourse [etc.], 1739), 227–43, in G. W. Prothero (ed.), *Select Statutes and Other Constitutional Documents Illustrative of the Reigns of Elizabeth and James I* (Oxford: The Clarendon Press, 1906), 286–93.

---

[2] The reference is to Phormion, a philosopher. He is said to have discoursed on the military arts before Hannibal, the great general, who regarded him as a blockhead.

[3] Approving.

stand taken for over forty years regarding its privileges, and because it sets forth ideas on necessary reforms.

To the King's most excellent Majesty, from the House of Commons assembled in Parliament. . . .

We know, and with great thankfulness to God acknowledge, that He hath given us a King of such understanding and wisdom as is rare to find in any prince in the world. Howbeit, seeing no human wisdom, how great soever, can pierce into the particularities of the rights and customs of people, or of the sayings and doings of particular persons, but by tract of experience and faithful report of such as know them . . . , what grief, what anguish of mind hath it been unto us at some times in presence to hear and see, in other things to find and feel by effect, your gracious Majesty, to the extreme prejudice of all your subjects of England and in particular of this House of the Commons thereof, so greatly wronged by misinformation . . . ! We have been constrained, as well in duty to your royal Majesty, whom with faithful hearts we serve, as to our dear native country, for which we serve in this Parliament . . . , freely to disclose unto your Majesty the truth of such matters concerning your subjects, the Commons, as hitherto by misinformation hath been suppressed or perverted. . . .

Now concerning the ancient right of the subjects of this realm, chiefly consisting in the privileges of this House of the Parliament, the misinformation openly delivered to your Majesty hath been in three things: first, that we hold not our privileges of right, but of grace only, renewed every parliament by way of donature[4] upon petition, and so to be limited; secondly, that we are no court of record, nor yet a court that can command view of records, but that our proceedings here are only to acts and memorials . . . ; thirdly and lastly, that the examination of the returns of writs for knights and burgesses[5] is without our compass, and due to the Chancery. Against which assertions . . . , tending directly and apparently to the utter overthrow of the very fundamental privileges of our house — and therein of the rights and liberties of the whole commons of your realm of England, which they and their ancestors from time immemorial have undoubtedly enjoyed under your Majesty's noble progenitors — we, the knights, citizens and burgesses of the House of Commons assembled in Parliament, and in the name of the whole commons of the realm of England, with uniform consent for ourselves and our posterity, do expressly protest, as being derogatory in the highest degree to the true dignity, liberty and authority of your Majesty's high courts of parliament,

[4] Gift.

[5] I.e., in connection with contested elections.

and consequently to the right of all your Majesty's said subjects, and the whole body of this your kingdom; and desire that this our protestation may be recorded to all posterity. And contrariwise . . . , we most truly avouch, first, that our privileges and liberties are our rights and due inheritance no less than our very lands and goods; secondly, that they cannot be withheld from us, denied or impaired but with apparent wrong to the whole state of the realm; thirdly, . . . that our making of request in the entrance of Parliament to enjoy our privilege is an act only of manners and doth not weaken our right, no more than our suing to the King for our lands by petition. . . .

What cause we, your poor Commons, have to watch over our privileges is manifest in itself to all men. The prerogatives of princes may easily and do daily grow; the privileges of the subject are for the most part at an everlasting stand. They may be by good providence and care preserved; but, being once lost, are not recovered but with much disquiet. If good kings were immortal as well as kingdoms, to strive so for privilege were but vanity perhaps and folly; but, seeing the same God, who in His great mercy hath given us a most wise King and religious, doth also sometimes permit hypocrites and tyrants in His displeasure and for the sins of people, from hence hath the desire of rights, liberties and privileges, both for nobles and commons, had his just original; by which an harmonical and stable state is framed, each member under the head enjoying that right and performing that duty which for the honor of the head and happiness of the whole is requisite. . . .

The rights of the liberties of the commons of England in Parliament consisteth chiefly in these three things: first, that the shires, cities and boroughs of England . . . have free choice of such persons as they shall put in trust to represent them; secondly, that the persons chosen, during the time of the Parliament as also of their access and recess, be free from restraint, arrest and imprisonment; thirdly, that in Parliament they may speak freely their consciences without check or controlment, doing the same with due reverence to the sovereign court of parliament – that is, to your Majesty and to both the houses, who all in this case make but one politic body, whereof your Highness is the head. . . .

For matter of religion, it will appear by examination of truth and right that your Majesty should be misinformed if any man should deliver that the kings of England have any absolute power in themselves, either to alter religion (which God defend should be in the power of any mortal man whatsoever!) or to make any laws concerning the same, otherwise than in temporal causes by consent of Parliament. We have and shall at all times by our oaths acknowledge that your Majesty is sovereign lord and supreme governor in both. Touching our own desires and proceedings therein, they have not been a little misconceived and misreported.

We have not come in any Puritan or Brownist[6] spirit to introduce their purity or to work the subversion of the state ecclesiastical as now it standeth — things so far and so clearly from our meaning as that, with uniform consent in the beginning of this Parliament, we committed to the Tower a man who out of that humor, in a petition exhibited to our house, had slandered the bishops. But . . . we come with another spirit, even with the spirit of peace. We disputed not of matters of faith and doctrine; our desire was peace only, and our device of unity: how this lamentable and long-lasting dissension amongst the ministers . . . might at length, before help come too late, be extinguished. And for the ways of this peace, we are not all addicted to our own inventions, but ready to embrace any fit way that may be offered. Neither desire we so much that any man, in regard of weakness of conscience, may be exempted after Parliament from obedience unto laws established, as that in this Parliament such laws may be enacted as, by relinquishment of some few ceremonies of small importance, or by any way better, a perpetual uniformity may be enjoined and observed. Our desire has been also to reform certain abuses crept into the ecclesiastical state, even as into the temporal; and lastly that the land might be furnished with a learned, religious and godly ministry, for the maintenance of whom we would have granted no small contribution, if in these, as we trust, just and religious desires we had found that correspondency from other which was expected. . . .

There remaineth . . . yet one part more of our duty at this present, which faithfulness of heart, not presumption, doth press. . . . Let your Majesty be pleased to receive public information from your Commons in Parliament as to the civil estate and government; for private informations pass often by practice. The voice of the people in the things of their knowledge is said to be as the voice of God. And if your Majesty shall vouchsafe at your best pleasure and leisure to enter into gracious consideration of our petition for the ease of those burdens under which your whole people have of long time mourned, hoping for relief by your Majesty, then may you be assured to be possessed of their hearts; and, if of their hearts, of all they can do or have.

And so we, your Majesty's most humble and loyal subjects, whose ancestors have with great loyalty, readiness and joyfulness served your famous progenitors, kings and queens of this realm, shall with like loyalty and joy, both we and our posterity serve your Majesty and your most royal issue forever with our lives, lands and goods, and all other our abilities; and by all means endeavor to procure your Majesty's honor with all plenty, tranquillity, content, joy and felicity.

[6] The reference is to Robert Browne's followers; see p. 208.

163. James and the Judges

James's attitude toward the law courts was that they were dependent upon the king. His point of view was well summed up by Sir Francis Bacon: "Let the judges be lions, but yet lions under the throne, being circumspect that they do not check or oppose any points of sovereignty."[7] In 1616 James warned the judges not to encroach upon the royal prerogative, or even deal with matters involving it without consultation with him or his Council.

Now having spoken of your office in general, I am next to come to the limits wherein you are bound yourselves, which . . . are three. First, encroach not upon the prerogative of the crown: if there falls out a question that concerns my prerogative or mystery of state, deal not with it, till you consult with the King or his Council, or both; for they are transcendent matters . . . That which concerns the mystery of the King's power is not lawful to be disputed; for that is to wade into the weakness of princes, and to take away the mystical reverence that belongs unto them that sit in the throne of God.

Secondly, that you keep yourselves within your own benches, not to invade other jurisdictions, which is unfit and an unlawful thing . . . Keep you therefore all in your own bounds, and for my part, I desire you to give me no more right, in my private prerogative, than you give to any subject, and therein I will be acquiescent; as for the absolute prerogative of the crown, that is no subject for the tongue of a lawyer, nor is lawful to be disputed.

It is atheism and blasphemy to dispute what God can do: good Christians content themselves with His will revealed in His word. So it is presumption and high contempt in a subject to dispute what a king can do, or say that a king cannot do this or that; but rest in that which is the king's revealed will in his law.

164. Coke's Defense of the Common Law

Since the judges ordinarily owed their appointments to the King, and held office at his discretion, it is to be expected that in most cases they would be compliant. Yet under James several cases involving issues in

---

SOURCES: *Works of James I* (London: R. Barker and J. Bill, 1616), 556, in G. W. Prothero (ed.), *Select Statutes and Other Constitutional Documents Illustrative of the Reigns of Elizabeth and James I* (Oxford: The Clarendon Press, 1906), 399–400.

Reports of Sir Edward Coke (London: J. Walthoe [etc.], 1727), pt. xii, 64–65.

---

[7] *Essays*, "Of Judicature."

dispute between King and Parliament resulted in decisions that ran counter to the King's claims. In this opposition Sir Edward Coke was particularly prominent. Coke held that the common law courts should, in general, maintain a position of independence between King and Parliament, exercising, as we should say, judicial review. The following remarks of Coke, made in 1607 when he was Chief Justice of Common Pleas, show his zealous support of common law procedures before majesty itself.

. . . Upon complaint made to him by Bancroft, Archbishop of Canterbury, concerning prohibitions,[8] the King was informed that, when the question was made of what matters the ecclesiastical judges have cognizance, either upon the expositions of the statutes concerning tithes or any other thing ecclesiastical, or upon the statute of 1 Elizabeth concerning the High Commission,[9] or in any other case in which there is not express authority in law, the King himself may decide it in his royal person; and that the judges are but the delegates of the King, and that the King may take what causes he shall please to determine from the determination of the judges and may determine them himself. And the Archbishop said that this was clear in divinity, that such authority belongs to the King by the word of God in the Scripture. To which it was answered by me, in the presence and with the clear consent of all the justices of England and barons of the Exchequer, that the King in his own person cannot adjudge any case, either criminal . . . or betwixt party and party . . . ; but this ought to be determined and adjudged in some court of justice according to the law and custom of England. . . .

Then the King said that he thought the law was founded upon reason, and that he and others had reason as well as the judges. To which it was answered by me that true it was that God had endowed his Majesty with excellent science and great endowments of nature; but his Majesty was not learned in the laws of his realm of England, and causes which concern the life or inheritance or goods or fortunes of his subjects, they are not to be decided by natural reason, but by the artificial reason and judgment of law — which law is an act which requires long study and experience, before that a man can attain to the cognizance of it — and that the law was the golden metwand[10] and measure to try the causes of the subjects, and which protected his Majesty in safety and peace. With which the King was greatly offended, and said that then he should be under the law — which was treason to affirm, as he said. To which I said

[8] Writs prohibiting courts from proceeding in a case.

[9] Established in 1559 to exercise the ecclesiastical jurisdiction conferred on the crown by the Act of Supremacy of that year.

[10] Measuring stick.

that Bracton saith *quod rex non debet esse sub homine, sed sub Deo et lege.*[11]

165. The Protestation of the Commons, 1621

In 1621 relations between James and the House of Commons stood at their lowest point of the reign. The Commons insisted that they were entitled to discuss public issues of all sorts, including the proposed marriage of Prince Charles with a Spanish princess. The King commanded them not to meddle with affairs of state. By way of rejoinder, the House voiced a protest, emphasizing its right to discuss such affairs freely and without molestation.

The Commons now assembled in Parliament, being justly occasioned thereunto concerning sundry liberties, franchises and privileges of Parliament, amongst others here mentioned, do make this protestation following:

That the liberties, franchises, privileges and jurisdictions of Parliament are the ancient and undoubted birthright and inheritance of the subjects of England; and that the arduous and urgent affairs concerning the King, state, and defense of the realm and of the Church of England, and the maintenance and making of laws, and redress of mischiefs and grievances which daily happen within this realm, are proper subjects and matter of counsel and debate in Parliament; and that in the handling and proceeding of these businesses every member of the House of Parliament hath, and of right ought to have, freedom of speech to propound, treat, reason, and bring to conclusion the same; and that the Commons in Parliament have like liberty and freedom to treat of these matters in such order as in their judgments shall seem fittest; and that every member of the said House hath like freedom from all impeachment, imprisonment, and molestation (other than by censure of the House itself) for or concerning any speaking, reasoning, or declaring of any matter or matters touching the Parliament or Parliament-business; and that if any of the said members be complained of and questioned for anything done or said in Parliament, the same is to be showed to the King by the advice and assent of all the Commons assembled in Parliament, before the King give credence to any private information.

SOURCE: John Rushworth, *Historical Collections* (London: J. A. for R. Boulter [etc.], 1680–1722), *I*, 53.

---

[11] The King should not be under man, but under God and the law.

SECTION B

Religious Friction under James I

166. The Millenary Petition, 1603

When James I ascended the English throne the Puritans had hopes that certain reforms in religion, not attained under Elizabeth, might be effected. With this in mind, they presented to him the Millenary Petition, so called because it was believed to carry at least a thousand signatures. The document is essentially moderate in tone, and refers to conditions long regarded as offensive in Puritan eyes.

Most gracious and dread Sovereign, seeing it hath pleased the Divine Majesty, to the great comfort of all good Christians, to advance your Highness, according to your just title, to the peaceable government of this church and commonwealth of England, we, the ministers of the Gospel in this land, neither as factious men affecting a popular parity in the church nor as schismatics aiming at the dissolution of the state ecclesiastical,[12] but as the faithful servants of Christ and loyal subjects of your Majesty, desiring and longing for the redress of divers abuses of the church, could do no less in our obedience to God, service to your Majesty, love to his church, than acquaint your princely Majesty with our particular griefs. . . . Our humble suit then unto your Majesty is that, [of] these offenses following, some may be removed, some amended, some qualified:

I. In the church service: that the cross in baptism, interrogatories ministered to infants, confirmation, as superfluous may be taken away. Baptism not to be ministered by women, and so explained. The cap and surplice not urged. That examination may go before the communion. That it be ministered with a sermon. That divers terms of priests and absolution and some other used, with the ring in marriage, and other such like in the book[13] may be corrected. The longsomeness of service abridged. Church songs and music moderated to better edification. That the Lord's day be not profaned, the rest upon holy days not so strictly urged. That there be an uniformity of doctrine prescribed. No popish opinion to be

---

SOURCE: Thomas Fuller, *Church History of Britain* (London: John Williams, 1655), bk. x, 22–23.

---

[12] The reference is to Presbyterian and Brownist objectives, which are disavowed.
[13] The Book of Common Prayer.

any more taught or defended, no ministers charged to teach their people to bow at the name of Jesus. That the canonical Scriptures only be read in the church.

II. Concerning church ministers: that none hereafter be admitted into the ministry but able and sufficient men, and those to preach diligently, and especially upon the Lord's day. That such as be already entered and cannot preach may either be removed and some charitable course taken with them for their relief, or else to be forced, according to the value of their livings, to maintain preachers. That non-residency be not permitted. That King Edward's statute for the lawfulness of ministers' marriage be revived. That ministers be not urged to subscribe but (according to the law) to the articles of religion and the King's supremacy only . . .

These, with such other abuses yet remaining and practiced in the Church of England, we are able to show not to be agreeable to the Scriptures, if it shall please your Highness further to hear us, or more at large by writing to be informed, or by conference among the learned to be resolved. And yet we doubt not but that without any further process your Majesty (of whose Christian judgment we have received so good a taste already) is able of yourself to judge of the equity of this cause. God, we trust, hath appointed your Highness our physician to heal these diseases. . . .

167. James's Antipathy to Presbyterianism

In answer to the Millenary Petition, James called a conference at the royal palace of Hampton Court, at which representatives of the Puritans were given an opportunity to debate with the bishops. Though the Petition aimed at no significant changes in church government, James got the impression from the debate that a changeover to Presbyterianism was being sought; as a result he denied virtually the whole petition, except for the request for a new version of the Scriptures in English. Like Elizabeth, James looked upon the tenets of Puritans and Separatists as dangerous to the monarchical authority.

Then he [Dr. Reynolds] desireth that according to certain provincial constitutions they of the clergy might have meetings once every three weeks; first in rural deaneries, and therein to have prophesying,[14] accord-

SOURCE: William Barlow, *The Sum and Substance of the Conference* (London: J. Windet for M. Law, 1604), 78f., in Edward P. Cheyney, *Readings in English History Drawns from the Original Sources* (Boston: Ginn, 1935), 430–31.

[14] Discussion of religious subjects.

ing as the Reverend Father Archbishop Grindal[15] and other bishops desired of her late Majesty; that such things as could not be resolved upon there might be referred to the archdeacon's visitation; and so from thence to the episcopal synod, where the bishop with his presbyters[16] should determine all such points as before could not be decided.

At which speech, his Majesty was somewhat stirred, yet, which is admirable in him, without passion or show thereof; thinking that they aimed at a Scottish presbytery, "which," saith he, "as well agreeth with a monarchy as God and the Devil. Then Jack and Tom and Will and Dick shall meet, and at their pleasure censure me and my Council and all our proceedings. Then Will shall stand up and say it must be thus; then Dick shall reply and say nay, marry, but we will have it thus. And therefore, here I must once reiterate my former speech, *le Roy s'avisera.*[17] Stay, I pray you, for one seven years, before you demand that of me, and if then you find me pursy[18] and fat and my windpipes stuffed, I will perhaps hearken to you; for let that government be once up, I am sure I shall be kept in breath; then shall we all of us have work enough, both our hands full. But, Doctor Reynolds, till you find that I grow lazy, let that alone.

"No bishop, no king, as before I said. Neither do I thus speak at random without ground, for I have observed since my coming into England that some preachers before me can be content to pray for James, King of England, Scotland, France, and Ireland, Defender of the Faith, but as for Supreme Governor in all causes and over all persons (as well ecclesiastical as civil) they pass that over with silence; and what cut they have been of, I after learned." After this, asking them if they had any more to object, and Dr. Reynolds answering no, his Majesty appointed the next Wednesday for both parties to meet before him, and rising from his chair, as he was going to his inner chamber, "If this be all," quoth he, "that they have to say, I shall make them conform themselves, or I will harry them out of the land, or else do worse.". . .

15 Edmund Grindal, Archbishop of Canterbury (1576–1582), was suspended because of his reluctance to enforce anti-Puritan legislation.

16 Priests; ministers below bishops and above deacons.

17 The king will consider it – a polite negative used in refusing assent to parliamentary legislation.

18 Short-winded.

SECTION C

Charles I and the Breakdown of Parliamentary Government

168. The Petition of Right, 1628

Charles I inherited his father's political convictions, but possessed less political aptitude. The relations between Crown and Parliament continued to be troubled. In 1628 the House of Commons, taking advantage of Charles's financial need, forced him to agree to the Petition of Right, the first important constitutional check on the powers of the Tudor-Stuart monarchy.

To the King's most excellent Majesty: Humbly show unto our sovereign lord the King, the Lords . . . and Commons in Parliament assembled that, whereas it is declared and enacted by a statute made in the time of the reign of King Edward the First, commonly called *Statutum de Tallagio non Concedendo*,[19] that no tallage or aid should be laid or levied by the King or his heirs in this realm without the goodwill and assent of the archbishops, bishops, earls, barons, knights, burgesses and other the freemen of the commonalty of this realm . . . : yet, nevertheless, of late divers commissions directed to sundry commissioners in several counties with instructions have issued, by means whereof your people have been in divers places assembled and required to lend certain sums of money unto your Majesty; and many of them, upon their refusal so to do, have had an oath administered unto them . . . and have been constrained to become bound to make appearance and give attendance before your Privy Council and in other places; and others of them have been therefor imprisoned, confined and sundry other ways molested and disquieted; and divers other charges have been laid and levied upon your people in several counties by lord lieutenants, deputy lieutenants, commissioners for musters, justices of peace and others, by command or direction from your Majesty or your Privy Council, against the laws and free customs of the realm.

And where also, by the statute called the Great Charter of the Liberties

---

SOURCE: *Statutes of the Realm* (London: Her Majesty's Stationery Office, 1810–28), V, 23–24.

---

[19] Actually, it seems, an imperfect abstract from the Confirmation of the Charters (1297); see p. 108.

of England,[20] it is declared and enacted that no freeman may be taken or imprisoned, or be disseised of his freehold or liberties or his free customs, or be outlawed or exiled or in any manner destroyed, but by the lawful judgment of his peers or by the law of the land . . . : nevertheless . . . divers of your subjects have of late been imprisoned without any cause showed; and when for their deliverance they were brought before your justices by your Majesty's writs of habeas corpus, there to undergo and receive as the court should order, and their keepers commanded to certify the causes of their detainer, no cause was certified but that they were detained by your Majesty's special command, signified by the lords of your Privy Council; and yet were returned back to several prisons without being charged with anything to which they might make answer according to the law.

And whereas of late great companies of soldiers and mariners have been dispersed into divers counties of the realm, and the inhabitants against their wills have been compelled to receive them into their houses, and there to suffer them to sojourn, against the laws and customs of this realm, and to the great grievance and vexation of the people; and whereas also by authority of Parliament in the five-and-twentieth year of the reign of King Edward III it is declared and enacted that no man should be forejudged of life or limb against the form of the Great Charter and the law of the land; and, by the said Great Charter and other the laws and statutes of this your realm, no man ought to be adjudged to death but by the laws established in this your realm, either by the customs of the same realm or by acts of Parliament; . . . nevertheless of late divers commissions under your Majesty's great seal have issued forth, by which certain persons have been assigned and appointed commissioners, with power and authority to proceed within the land according to the justice of martial law against such soldiers or mariners, or other dissolute persons joining with them, as should commit any murder, robbery, felony, mutiny or other outrage or misdemeanor whatsoever, and by such summary course and order as is agreeable to martial law and as is used in armies in time of war to proceed to the trial and condemnation of such offenders, and them to cause to be executed and put to death according to the law martial; . . . which commissions and all other of like nature are wholly and directly contrary to the said laws and statutes of this your realm.

They do therefore humbly pray your most excellent Majesty that no man hereafter be compelled to make or yield any gift, loan, benevolence, tax or such like charge without common consent by act of Parliament; and that none be called to make answer, or take such oath, or to give attendance, or be confined or otherwise molested or disquieted concerning

20 Magna Carta (1215). The reference is to the thirty-ninth article; see p. 82.

the same, or for refusal thereof; and that no freeman, in any such manner as is before mentioned, be imprisoned or detained; and that your Majesty would be pleased to remove the said soldiers and mariners; and that your people may not be so burdened in time to come; and that the aforesaid commissions for proceeding by martial law may be revoked and annulled; and that hereafter no commissions of like nature may issue forth to any person or persons whatsoever, to be executed as aforesaid, lest by color of them any of your Majesty's subjects be destroyed or put to death, contrary to the laws and franchise of the land.

All which they most humbly pray of your most excellent Majesty as their rights and liberties according to the laws and statutes of this realm; and that your Majesty would also vouchsafe to declare that the awards, doings and proceedings to the prejudice of your people in any of the premises shall not be drawn hereafter into consequence or example; and that your Majesty would be also graciously pleased, for the further comfort and safety of your people, to declare your royal will and pleasure that in the things aforesaid all your officers and ministers shall serve you according to the laws and statutes of this realm, as they tender the honor of your Majesty and the prosperity of this kingdom.

169. Charles I on the Petition of Right, 1628

The acceptance of the Petition of Right was followed immediately by wrangling between King and Commons regarding the extent to which that document limited the royal taxing powers, specifically in connection with the import duties known as tunnage and poundage, which for two hundred years had been regarded as part of the royal revenue. In proroguing Parliament in 1628, Charles showed that he had not abandoned the divine right theory of kingship, and explicitly pointed out that the Petition did not extend to these duties.

My Lords and Gentlemen:

It may seem strange that I come so suddenly to end this session. Wherefore, before I give my assent to the bills, I will tell you the cause; though I must avow that I owe an account of my actions to none but God alone. It is known to everyone that a while ago the House of Commons gave me a remonstrance,[21] how acceptable every man may judge, and for the merit

SOURCE: *Journals of the House of Lords* (London: Her Majesty's Stationery Office), *III*, 879.

[21] The Petition of Right (1628); see p. 248.

of it, I will not call that in question; for I am sure no wise man can justify it.

Now, since I am certainly informed that a second remonstrance is preparing for me, to take away my profit of tunnage and poundage (one of the chief maintenances of the Crown) by alleging that I have given away my right thereof by my answer to your petition, this is so prejudicial unto me that I am forced to end this session some few hours before I meant it, being willing not to receive any more remonstrances to which I must give an harsh answer. And since I see that even the House of Commons begins already to make false constructions of what I granted in your petition; lest it might be worse interpreted in the country, I will now make a declaration concerning the true meaning thereof.

The profession in both Houses, in time of hammering this petition, was by no ways to entrench upon my prerogative, saying they had neither intention nor power to hurt it. Therefore it must needs be conceived that I have granted no new, but only confirmed the ancient, liberties of my subjects. Yet, to show the clearness of my intentions, that I neither repent nor mean to recede from anything I have promised you, I do here declare that those things which have been done – whereby men had some cause to suspect the liberty of the subjects to be trenched upon, which indeed was the first and true ground of the petition – shall not hereafter be drawn into example for your prejudice; and in time to come – in the word of a king – you shall not have the like cause to complain.

But as for tunnage and poundage, it is a thing I cannot want[22] and was never intended by you to ask, never meant, I am sure by me to grant. To conclude, I command you all that are here to take notice of what I have spoken at this time, to be the true intent and meaning of what I granted you in your petition. . . .

170. The Resolutions of the Commons, 1629

When Charles's third Parliament reassembled for its second session in 1629, the Commons immediately began to debate the controversial subject of tunnage and poundage, as well as the even more controversial issue of religion. Many of the members were concerned over the innovations of the anti-Calvinistic wing of the Church, led by Laud, which the King strongly supported. Charles speedily called for dissolution, but before it was carried out the Commons, in defiance of the royal order, framed these three resolutions.

SOURCE: John Rushworth, *Historical Collections* (London: J. A. for R. Boulter, 1680–1722), *I*, 660.

---

[22] Be without.

Whosoever shall bring in innovation of religion, or by favor or countenance seek to extend or introduce popery or Arminianism,[23] or other opinion disagreeing from the true and orthodox Church, shall be reputed a capital enemy to this kingdom and commonwealth.

Whosoever shall counsel or advise the taking and levying of the subsidies of tunnage and poundage, not being granted by Parliament, or shall be an actor or instrument therein, shall be likewise reputed an innovator in the government, and a capital enemy to the kingdom and commonwealth.

If any merchant or person whatsoever shall voluntarily yield or pay the said subsidies of tunnage and poundage, not being granted by Parliament, he shall likewise be reputed a betrayer of the liberties of England, and an enemy to the same.

171. The Judges' Opinion Regarding Ship Money, 1637

Between 1629 and 1640 the legislators were never called into session. Without Parliament to grant supply, Charles was forced to raise funds in other ways. The most unpopular tax device was ship money, an ancient levy on coastal communities which was now extended throughout the realm. Aware of opposition, Charles early in 1637 sounded out the judges, and received a favorable opinion. This, as given below, was published. Even so, John Hampden was to question the tax in the courts the following year.

We,[24] desirous to avoid . . . inconveniences, and out of our princely love and affection to all our subjects, being willing to prevent such errors as any of our loving subjects may happen to run into, have thought fit in a case of this nature[25] to advise with our judges, who, we doubt not, are all well studied and informed in the right of our sovereignty. And because the trial in our several courts, by the formality in pleading, will require a long protraction, we have thought it expedient, by this our letter directed to you all, to require your judgments in the case, as it is set down in the enclosed paper, which will not only gain time but also

SOURCE: Thomas B. Howell, *Complete Collection of State Trials* (London: R. Bagshaw [etc.], 1809–26), *III*, 843–44.

[23] The Arminians (followers of Jacob Harmensen, whose Latinized name was Arminius) opposed some Calvinistic doctrines and supported the right of the state to control the church.

[24] Charles I here gives his instructions to the judges.

[25] I.e., a case involving the legality of ship money.

be of more authority to overrule any prejudicate[26] opinions of others in the point . . .

May it please your most excellent Majesty:

We have, according to your Majesty's command – every man by himself and all of us together – taken into consideration the case and question signed by your Majesty and enclosed in your royal letter. And we are of opinion that, when the good and safety of the kingdom in general is concerned, and the whole kingdom in danger, your Majesty may, by writ under the great seal of England, command all the subjects of this your kingdom at their charge to provide and furnish such number of ships, with men, munition and victuals, and for such time as your Majesty shall think fit, for the defense and safeguard of the kingdom from such danger and peril; and that by law your Majesty may compel the doing thereof in case of refusal or refractoriness. And we are also of opinion that in such case your Majesty is the sole judge both of the danger and when and how the same is to be prevented and avoided. . . .

SECTION D

From Remonstrance to Rebellion

172. The Grand Remonstrance, 1641

In 1640 Charles I found it necessary to call Parliament in order to secure funds to proceed against the Scots, who had rebelled as a result of an attempt to force Anglican religious practices upon them, and invaded northern England. Taking advantage of the King's difficulties, Parliament carried through a program of legislation designed to limit the royal power; thus Parliament was to meet at least every three years, and arbitrary courts and ship money were abolished. These measures were enacted with remarkable unanimity. But the vote on the Grand Remonstrance, presented to Charles by the Commons late in 1641, reveals an almost even balance between royalist and reforming interests. The Remonstrance, sponsored by the more radical Members, summarized grievances against the King's government, commented on what Parliament had done to redress them, and pushed for further reforms.

SOURCE: John Rushworth, *Historical Collections* (London: J. A. for R. Boulter, 1680–1722), *IV*, 438.

---

[26] Prematurely conceived.

The principal objectives are to be found in this petition accompanying the Remonstrance:

We, your most humble and obedient subjects, do with all faithfulness and humility beseech your Majesty:

1. That you will be graciously pleased to concur with the humble desires of your people in a parliamentary way, for the preserving the peace and safety of the kingdom from the malicious designs of the popish party;

For depriving the bishops of their votes in Parliament, and abridging their immoderate power usurped over the clergy and other your good subjects, which they have perniciously abused to the hazard of religion and great prejudice and oppression of the laws of the kingdom and just liberty of your people;

For the taking away such oppressions in religion, church government and discipline as have been brought in and fomented by them;

For uniting all such your loyal subjects together as join in the same fundamental truths against the papists, by removing some oppressions and unnecessary ceremonies by which divers weak consciences have been scrupled, and seem to be divided from the rest, and for the due execution of those good laws which have been made for securing the liberty of your subjects.

2. That your Majesty will likewise be pleased to remove from your Council all such as persist to favor and promote any of those pressures and corruptions wherewith your people have been grieved, and that for the future your Majesty will vouchsafe to employ such persons in your great and public affairs, and to take such to be near you in places of trust, as your Parliament may have cause to confide in; that in your princely goodness to your people you will reject and refuse all mediation and solicitation to the contrary, how powerful and near soever.

3. That you will be pleased to forbear to alienate any of the forfeited and escheated lands in Ireland which shall accrue to your crown by reason of this rebellion,[27] that out of them the crown may be the better supported, and some satisfaction made to your subjects of this kingdom for the great expenses they are like to undergo [in] this war.

Which humble desires of ours being graciously fulfilled by your Majesty, we will, by the blessing and favor of God, most cheerfully undergo the hazard and expenses of this war, and apply ourselves to such other courses and counsels as may support your royal estate with honor and plenty at home, with power and reputation abroad, and by our loyal affections, obedience and service lay a sure and lasting foundation of the

[27] An Irish rebellion broke out in 1641, beginning with a rising in Ulster against English Protestant settlers.

greatness and prosperity of your Majesty, and your royal prosperity in future times.

173. Charles I's Answer to the Petition in the Grand Remonstrance

Displeased with the Grand Remonstrance, and especially by its publication as a piece of political propaganda, Charles voiced his reactions in the following message.

We having received from you, soon after our return out of Scotland, a long petition consisting of many desires of great moment, together with a declaration of a very unusual nature annexed thereunto, we had taken some time to consider of it, as befitted us in a matter of that consequence, being confident that your own reason and regard to us, as well as our express intimation by our Comptroller to that purpose, would have restrained you from the publishing of it till such time as you should have received our answer to it; but, much against our expectations, finding the contrary, that the said declaration is already abroad in print, by directions from your House as appears from the printed copy, we must let you know that we are very sensible of the disrespect. Notwithstanding, it is our intention that no failing on your part shall make us fail in ours of giving all due satisfaction to the desires of our people in a parliamentary way; and therefore we send you this answer to your petition, reserving ourself in point of the declaration which we think unparliamentary, and shall take a course to do that which we shall think fit in prudence and honor.

To the petition, we say that although there are divers things in the preamble of it which we are so far from admitting that we profess we cannot at all understand them, . . . so that the prayers of your petition are grounded upon such premises as we must in no wise admit; yet, notwithstanding, we are pleased to give this answer to you.

To the first, concerning religion, consisting of several branches, we say that, for preserving the peace and safety of this kingdom from the design of the popish party, we have, and will still, concur with all the just desires of our people in a parliamentary way; [but] that, for the depriving of the bishops of their votes in Parliament, we would have you consider that their right is grounded upon the fundamental law of the kingdom and constitution of Parliament. This we would have you consider; but since you desire our concurrence in a parliamentary way, we will give no further answer at this time.

SOURCE: John Rushworth, *Historical Collections* (London: J. A. for R. Boulter, 1680–1722), *IV*, 452–53.

As for the abridging of the inordinate power of the clergy, we conceive that the taking away of the High Commission Court[28] hath well moderated that; but if there continue any usurpations or excesses in their jurisdictions, we therein neither have nor will protect them.

Unto that clause which concerneth corruptions (as you style them) in religion, in church government and in discipline, and the removing of such unnecessary ceremonies as weak consciences might check at: [we reply] that for any illegal innovations which may have crept in, we shall willingly concur in the removal of them; that if our Parliament shall advise us to call a national synod, which may duly examine such ceremonies as give just course of offence to any, we shall take it into consideration and apply ourself to give due satisfaction therein; but we are very sorry to hear, in such general terms, corruption in religion objected, since we are persuaded in our consciences that no church can be found upon the earth that professeth the true religion with more purity of doctrine than the Church of England doth, nor where the government and discipline are jointly more beautiful and free from superstition than as they are here established by law, which, by the grace of God, we will with constancy maintain (while we live) in their purity and glory, not only against all invasions of popery but also from the irreverence of those many schismatics and separatists, wherewith of late this kingdom and this city abounds, to the great dishonor and hazard both of church and state, for the suppression of whom we require your timely aid and active assistance.

To the second prayer of the petition, concerning the removal and choice of councillors, we know not any of our Council to whom the character set forth in the petition can belong: that by those whom we had exposed to trial, we have already given you sufficient testimony that there is no man so near unto us in place or affection whom we will not leave to the justice of the law, if you shall bring a particular charge and sufficient proofs against him; and of this we do again assure you, but in the meantime we wish you to forbear such general aspersions as may reflect upon all our Council, since you name none in particular.

That for the choice of our councillors and ministers of state, it were to debar us that natural liberty all freemen have; and as it is the undoubted right of the crown of England to call such persons to our secret counsels, to public employment and our particular service as we shall think fit, so we are and ever shall be very careful to make election of such persons in those places of trust as shall have given good testimonies of their abilities and integrity, and against whom there can be no just cause of

[28] See p. 243. This court, the principal tribunal in matters of church discipline, had been abolished, along with other arbitrary courts, by statute earlier in 1641.

exception whereon reasonably to ground a diffidence;[29] and to choices of this nature we assure you that the mediation of the nearest unto us hath always concurred.

To the third prayer of your petition concerning Ireland, we understand your desire of not alienating the forfeited lands thereof to proceed from much care and love, and likewise that it may be a resolution very fit for us to take; but whether it be seasonable to declare resolutions of that nature before the events of a war be seen, that we much doubt of. Howsoever, we cannot but thank you for this care and your cheerful engagement for the suppression of that rebellion, upon the speedy effecting whereof the glory of God in the Protestant profession, the safety of the British there, our honor and that of the nation, so much depends; all the interests of this kingdom being so involved in that business, we cannot but quicken your affections therein, and shall desire you to frame your counsels to give such expedition to the work as the nature thereof and the pressures in point of time require, and whereof you are put in mind by the daily insolence and increase of those rebels.

For conclusion, your promise to apply yourselves to such courses as may support our royal estate with honor and plenty at home and with power and reputation abroad is that which we have ever promised ourself, both from your loyalties and affections, and also for what we have already done, and shall daily go adding unto, for the comfort and happiness of our people.

174. The Nineteen Propositions, 1642

Late in the summer of 1641 an Irish rebellion raised a new constitutional issue: whether the military forces required to quell it should be controlled by King or Parliament. In 1642 Parliament framed the Militia Ordinance, under which it assumed such control. When the King refused to accept it, Parliament declared it to be the law of the land, even without royal sanction. This breakdown of relations was never mended. Charles withdrew from his capital and refused to communicate with the legislature, which exercised executive powers and issued ordinances as the law of the land. Late in the spring of 1642 Parliament attempted to regularize and extend its new position in the Nineteen Propositions, the total effect of which (had Charles accepted them) would have been to make the king a figurehead.

SOURCE: *Journals of the House of Lords* (London: Her Majesty's Stationery Office), V, 97–98.

---

[29] Distrust.

Your Majesty's most humble and faithful subjects, the Lords and Commons in Parliament, having nothing in their thoughts and desires more precious and of higher esteem — next to the honor and immediate service of God — than the just and faithful performance of their duty to your Majesty and this kingdom . . . , do in all humility and sincerity present to your Majesty their most dutiful petition and advice, that out of your princely wisdom . . . you will be pleased to grant and accept these their humbles desires and propositions, as the most necessary effectual means . . . of removing those jealousies and differences which have unhappily fallen betwixt you and your people . . . :

1. First, that the lords and others of your Majesty's Privy Council and such great officers and ministers of state, either at home or beyond the seas, may be put from your Privy Council and from those offices and employments, excepting such as shall be approved of by both houses of Parliament; and that the persons put into the places and employments of those that are removed may be approved of by both houses of Parliament; and that all Privy Councillors shall take an oath for the due execution of their places in such form as shall be agreed upon by both houses of Parliament.

2. That the great affairs of the kingdom may not be concluded or transacted by the advice of private men, or by any unknown or unsworn councillors; but that such matters as concern the public and are proper for the High Court of Parliament, which is your Majesty's great and supreme council, may be debated, resolved and transacted only in Parliament, and not elsewhere. And such as shall presume to do anything to the contrary shall be reserved to the censure and judgment of Parliament. And such other matters of state as are proper for your Majesty's Privy Council shall be debated and concluded by such of the nobility and others as shall from time to time be chosen for that place, by approbation of both houses of Parliament. And that no public act concerning the affairs of the kingdom, which are proper for your Privy Council, may be esteemed of any validity, as proceeding from the royal authority, unless it be done by the advice and consent of the major part of your Council, attested under their hands. And that your Council may be limited to a certain number, not exceeding twenty-five, nor under fifteen. . . .

3. That the Lord High Steward of England, Lord High Constable, Lord Chancellor or Lord Keeper of the Great Seal, Lord Treasurer, Lord Privy Seal, Earl Marshal, Lord Admiral, Warden of the Cinque Ports,[30] Chief Governor of Ireland, Chancellor of the Exchequer, Master of the

[30] An ancient office associated with certain towns (originally five) in southeastern England, which bore special obligations in providing for the fleet.

Wards, Secretaries of State, two Chief Justices and Chief Baron[31] may always be chosen with the approbation of both houses of Parliament. . . .

4. That he or they unto whom the government and education of the King's children shall be committed shall be approved of by both houses of Parliament.

5. That no marriage shall be concluded or treated for any of the King's children with any foreign prince or other person whatsoever, abroad or at home, without the consent of Parliament. . . .

6. That the laws in force against Jesuits, priests and popish recusants be strictly put in execution, without any toleration or dispensation to the contrary. . . .

7. That the votes of popish lords in the house of peers may be taken away, so long as they continue papists . . .

8. That your Majesty would be pleased to consent that such a reformation be made of the church government and liturgy as both houses of Parliament shall advise. . . .

9. That your Majesty will be pleased to rest satisfied with that course that the Lords and Commons have appointed for ordering the militia until the same shall be further settled by a bill; and that your Majesty will recall your declarations and proclamations against the ordinance made by the Lords and Commons concerning it.

10. That such members of either house of Parliament as have, during this present Parliament, been put out of any place and office may either be restored to that place and office or otherwise have satisfaction for the same, upon the petition of that house whereof he or they are members.

11. That all privy councillors and judges may take an oath, the form whereof to be agreed on and settled by act of Parliament, for the maintaining of the Petition of Right and of certain statutes made by this Parliament, which shall be mentioned by both houses of Parliament. . . .

12. That all the judges and all the officers placed by approbation of both houses of Parliament may hold their places *quam diu bene se gesserint.*[32]

13. That the justice of Parliament may pass upon all delinquents, whether they be within the kingdom or fled out of it. . . .

14. That the general pardon offered by your Majesty may be granted, with such exceptions as shall be advised by both houses of Parliament.

15. That the forts and castles of this kingdom may be put under the command and custody of such persons as your Majesty shall appoint with the approbation of your Parliament. . . .

16. That the extraordinary guards and military forces now attending your Majesty may be removed and discharged; and that for the future

31 The presiding judge in the Court of Exchequer.
32 During good behavior.

you will raise no such guards or extraordinary forces, but according to the law, in case of actual rebellion or invasion.

17. That your Majesty will be pleased to enter into a more strict alliance with the states of the United Provinces, and other neighboring princes and states of the Protestant religion, for the defense and maintenance thereof . . .

18. That your Majesty will be pleased, by act of Parliament, to clear the Lord Kimbolton and the five members of the House of Commons[33] in such manner that future parliaments may be secured from the consequence of that evil precedent.

19. That your Majesty will be graciously pleased to pass a bill for restraining peers made hereafter from sitting or voting in Parliament unless they be admitted thereunto with the consent of both houses of Parliament.

And these our humble desires being granted by your Majesty, we shall forthwith apply ourselves to regulate your present revenue in such sort as may be for your best advantage; and likewise to settle such an ordinary and constant increase of it as shall be sufficient to support your royal dignity in honor and plenty, beyond the proportion of any former grants of the subjects of this kingdom to your Majesty's royal predecessors . . . , and cheerfully employ to the uttermost our power and endeavor in the real expression and performance of our most dutiful and loyal affections, to the preserving and maintaining the royal honor, greatness and safety of your Majesty and your posterity.

175. Charles I's Answer to the Nineteen Propositions

In rejecting the Nineteen Propositions, Charles I issued a statement emphasizing the advantages of mixed government, as long as a balance could be maintained between king, lords and commons. With such a balance tyranny, in Charles's opinion, would be impossible. But if the Nineteen Propositions were adopted the balance would be destroyed, and chaos would result. On this occasion, understandably enough, Charles said nothing about divine right. The theory of mixed monarchy later came to be regarded as basic to the constitution, particularly in the century and a half following the Revolution of 1688.

SOURCE: John Rushworth, *Historical Collections* (London: J. A. for R. Boulter, 1680–1722), *IV*, 731–32.

---

[33] King Charles had attempted to arrest them, on grounds of treasonable correspondence with the Scots, earlier in 1642.

There being three kinds of government among men, absolute monarchy, aristocracy and democracy, and all these having their particular conveniences and inconveniences. The experience and wisdom of your ancestors hath so moulded this [present government] out of a mixture of these, as to give to this kingdom (as far as human prudence can provide) the conveniences of all three, without the inconveniences of any one, as long as the balance hangs even between the three estates, and they run jointly on in their proper channel . . . and the overflowing of either on either side raise no deluge or inundation. The ill of absolute monarchy is tyranny; the ill of aristocracy is faction and division; the ills of democracy are tumults, violence and licentiousness. The good of monarchy is the uniting a nation under one head, to resist invasion from abroad and insurrection at home; the good of aristocracy is the conjunction of counsel in the ablest persons of a state for the public benefit: the good of democracy is liberty, and the courage and industry which liberty begets.

In this kingdom the laws are jointly made by a king, by a house of peers, and by a house of commons chosen by the people, all having free votes and particular privileges. The government according to these laws is entrusted to the king: power of treaties of war and peace, of making peers, of choosing officers and counsellors for state, judges for law, commanders for forts and castles; giving commissions for raising men; to make war abroad, or to prevent or provide against invasions or insurrections at home; benefit of confiscations, power of pardoning, and some more of the like kind are placed in the king. And this kind of regulated monarchy, having this power to preserve that authority, without which it would be disabled to preserve the laws in their force and the subjects in their liberties and properties, is intended to draw to him such a respect and relation from the great ones as may hinder the ills of division and faction; and such a fear and reverence from the people as may hinder tumults, violence and licentiousness.

Again, that the prince may not make use of this high and perpetual power to the hurt of those for whose good he hath it, and make use of the name of public necessity for the gain of his private fortunes and followers, to the detriment of his people, the House of Commons (an excellent conserver of liberty, but never intended for any share in government or the choosing of them that should govern) is solely entrusted with the first propositions concerning the levies of moneys . . . and the impeaching of those, who for their own ends, though countenanced by any surreptitiously gotten command of the king, have violated that law, which he is bound (when he knows it) to protect; and to the prosecution of which they were bound to advise him, at least not to serve him in the contrary. And the Lords being trusted with a judicatory power are an excellent screen and blank between the prince and people, to assist each against any encroachments of the other, and by just judgments to preserve that law which ought to be the rule of every one of the three . . .

Since, therefore, the power legally placed in both houses is more than sufficient to prevent and restrain the power of tyranny, and [since] without the power which is now asked from us we shall not be able to discharge that trust which is the end of monarchy; since this would be a total subversion of the fundamental laws and that excellent constitution of this kingdom . . . ; since to the power of punishing (which is already in your hands according to law) if the power of preferring be added, we shall have nothing left to us but to look on; since the encroaching of one of these estates upon the power of the other is unhappy in the effects, both to them and all the rest . . . : for all these reasons to all these demands our answer is *nolumus leges Angliae mutari,*[34] but this we promise, that we will be as careful of preserving the laws in what is supposed to concern wholly our subjects as in what most concerns ourself. . . .

176. The Solemn League and Covenant, 1643

The Civil War began in August, 1642. At first fortune generally favored the King's forces. This situation led in 1643 to an agreement with the Scots, whereby they consented to assist Parliament with an army, in return for financial support and the establishment of Presbyterianism in England. This engagement is known as the Solemn League and Covenant. It is noteworthy that the allies included a pledge to preserve the King's "just power and greatness." Almost no one in England favored the overthrow of the monarchy at this time; the Scots were always overwhelmingly against it.

We, noblemen, barons, knights, gentlemen, citizens, burgesses, ministers of the Gospel, and commons of all sorts in the kingdoms of England, Scotland and Ireland, . . . have now at last . . . for the preservation of ourselves and our religion from utter ruin and destruction, according to the commendable practice of these kingdoms in former times, and the example of God's people in other nations, after mature deliberation resolved and determined to enter into a mutual and solemn league and covenant, wherein we all subscribe; and . . . do swear:

That we shall sincerely, really and constantly, through the grace of God, endeavor in our several places and callings the preservation of the

SOURCE: John Rushworth, *Historical Collections* (London: J. A. for R. Boulter, 1680–1722), V, 478–79.

[34] We do not wish the laws of England to be changed.

reformed religion in the Church of Scotland . . . ; the reformation of religion in the kingdoms of England and Ireland . . . ; and we shall endeavor to bring the churches of God in the three kingdoms to the nearest conjunction and uniformity in religion, confession of faith, form of church government, directory for worship and catechizing . . .

That we shall . . . endeavor the extirpation of popery, prelacy . . . , superstition, heresy, schism, profaneness and whatsoever shall be found to be contrary to sound doctrine and power of godliness . . .

We shall . . . endeavor with our estates and lives mutually to preserve the rights and privileges of the parliaments, and the liberties of the kingdoms, and to preserve and defend the King's Majesty's person and authority, in the preservation and defense of the true religion and liberties of the kingdoms, that the world may bear witness with our consciences of our loyalty, and that we have no thoughts or intentions to diminish his Majesty's just power and greatness. . . .

And whereas the happiness of a blessed peace between these kingdoms . . . is by the good providence of God granted to us, and hath been lately concluded and settled by both parliaments, we shall . . . endeavor that they may remain conjoined in a firm peace and union to all posterity, and that justice may be done upon the wilful opposers thereof. . . .

SECTION E

The Fall of the Monarchy

177. A Debate in the Army, 1647

The defeat of the King in 1646 brought an opportunity for reconstruction of church and state. But the victors did not see eye to eye. Presbyterians differed from Independents, and Parliament from the Army, which remained in being and was itself split between radical and conservative camps. Debates were held among the soldiers which are of particular interest in that they reveal a strong current of support for democracy and religious toleration among the rank and file. The following extracts are from a debate at Putney late in 1647. Under discussion was an article of the proposed "Agreement of the People,"

SOURCE: C. H. Firth (ed.), *Selections from the Papers of William Clarke* (London: Royal Historical Society, 1891–1901), *I*, 301–29.

which called for representation "more indifferently [impartially] proportioned according to the number of the inhabitants."

COL. RAINBOROUGH.[35]. . . I think that the poorest he that is in England hath a life to live as the greatest he; and therefore, truly, Sir, I think it's clear that every man that is to live under a government ought first by his own consent to put himself under that government . . .

COMMISSARY IRETON.[36] Give me leave to tell you that if you make this the rule, I think you must fly for refuge to an absolute natural right, and you must deny all civil right. . . . For my part I think it is no right at all. I think that no person hath a right to an interest or share in the disposing or determining of the affairs of the kingdom, and in choosing those that shall determine what laws we shall be ruled by here, no person hath a right to this that hath not a permanent fixed interest in this kingdom . . . We talk of birthright. Truly [by] birthright there is thus much claim. Men may justly have by birthright, by their very being born in England, that we should not seclude them out of England, that we should not refuse to give them air and place and ground and the freedom of the highways and other things, to live amongst us. . . . That I think is due to a man by birth. But that by a man's being born here he shall have a share in that power that shall dispose of the lands here, and of all things here, I do not think it a sufficient ground. I am sure if we look upon . . . that which is most radical[37] and fundamental and which if you take away there is no man hath any land, any goods, any civil interest, that is this: that those that choose the representers[38] for the making of laws by which this state and kingdom are to be governed are the persons who, taken together, do comprehend the local interest of this kingdom; that is, the persons in whom all land lies, and those in corporations in whom all trading lies . . .

RAINBOROUGH. Truly, Sir, I am of the same opinion I was, and am resolved to keep it till I know reason why I should not. . . . I do think that the main cause why almighty God gave men reason, it was that they should make use of that reason. . . . And truly I think that half a loaf is better than none if a man be an hungry, yet I think there is nothing that God hath given a man that any [one] else can take from him. . . . I do not find anything in the law of God that a lord shall choose twenty burgesses and a gentleman but two, or a poor man shall choose none. . . .

[35] Thomas Rainborough, leader of the republican wing of the officers.

[36] Henry Ireton, Quartermaster General. Like Oliver Cromwell, his father-in-law, he for a time sought to maintain the monarchy. Later both men were leaders of the regicides.

[37] Pertaining to the root of the matter; basic.

[38] Representatives.

But I do find that all Englishmen must be subject to English laws, and I do verily believe that there is no man but will say that the foundation of all law lies in the people . . .

IRETON. . . . I wish we may all consider of what right you will challenge, that all the people should have right to elections. Is it by the right of nature? If you will hold forth that as your ground, then I think you must deny all property too, and this is my reason. For thus: by that same right of nature . . . by which you can say one man hath an equal right with another to the choosing of him that shall govern him – by the same right of nature he hath an equal right in any goods he sees: meat, drink, clothes, to take and use them for his sustenance. He hath a freedom to the land, [to take] the ground, to exercise it, till it; he hath the [same] freedom to anything that anyone doth account himself to have any propriety[39] in . . . Since you cannot plead to it by anything but the law of nature, . . . I would fain have any man show me their bounds, where you will end, and [why you should not] take away all property?

RAINBOROUGH. . . . I wish we were all true hearted, and that we did all carry ourselves with integrity. . . . For my part, as I think, you forgot something that was in my speech, and you do not only yourselves believe that [we] are inclining to anarchy, but you would make all men believe that. . . . That there's a property the law of God says it; else why [hath] God made that law, "Thou shalt not steal"? . . . I wish you would not make the world believe that we are for anarchy.

CROMWELL. I know nothing but this, that they that are the most yielding have the greatest wisdom; but really, Sir, this is not right as it should be. No man says that you have a mind to anarchy, but the consequence of this rule tends to anarchy, must end in anarchy; for where is there any bound or limit set, if you take away this [limit] that men that have no interest but the interest of breathing [shall have no voice in elections]? Therefore I am confident . . . we should not be so hot one with another.

RAINBOROUGH. . . . I deny that there is a property, to a lord, to a gentleman, to any man more than another in the kingdom of England. . . . I would fain know what we have fought for, and this is the old law of England and that which enslaves the people of England, that they should be bound by laws in which they have no voice at all . . .

MR. SEXBY.[40]. . . We have engaged in this kingdom and ventured our lives, and it was all for this: to recover our birthrights and privileges as Englishmen; and by the arguments used there is none. There are many thousands of us soldiers that have ventured our lives; we have had little propriety in the kingdom as to our estates, yet we have had a birthright.

[39] Property.
[40] Captain Edward Sexby, a prominent Leveller. He later published a defense of tyrannicide.

But it seems now, except a man hath a fixed estate in this kingdom, he hath no right in this kingdom. I wonder we were so much deceived. If we had not a right to the kingdom we were mere mercenary soldiers. . . . I shall tell you in a word my resolution. I am resolved to give my birthright to none. . . . I do think the poor and meaner of this kingdom . . . have been the means of the preservation of this kingdom. . . .

IRETON. I am very sorry we are come to this point, that from reasoning one to another we should come to express our resolutions. . . . For my part, rather than I will make a disturbance to a good constitution of a kingdom wherein I may live in godliness and honesty and peace and quietness, I will part with a great deal of my birthright. I will part with my own property rather than I will be the man that shall make a disturbance in the kingdom for my property. . . .

RAINBOROUGH. . . . But I would fain know what the soldier hath fought for all this while? He hath fought to enslave himself, to give power to men of riches, men of estates, to make him a perpetual slave. We do find in all presses[41] that go forth none must be pressed that are freehold men. When these gentlemen fall out among themselves they shall press the poor scrubs[42] to come and kill them. . . .

CROMWELL. I confess I was most dissatisfied with that I heard Mr. Sexby speak of any man here, because it did savor so much of will. But I desire that all of us may decline that, and if we meet here really to agree to that which was for the safety of the kingdom, let us not spend so much time in such debates as these are, but let us apply ourselves to such things as are conclusive, and that shall be this: everybody here would be willing that the representative might be mended, that is, it might be better than it is . . . I say it again, if I cannot be satisfied to go so far as these gentlemen that bring this paper, I profess I shall freely and willingly withdraw myself, and I hope to do it in such manner that the army shall see that I shall by my withdrawing satisfy the interest of the army, the public interest of the kingdom, and those ends these men aim at.

178. An Agreement of the People, 1649

Originally drawn up in 1647, this proposed constitution was presented, with some modifications, to the House of Commons in January, 1649. Emanating from the Levellers, it appealed to the people as the source

SOURCE: William Cobbett, *Parliamentary History of England* (London: R. Bagshaw [etc.], 1806–20), *III* 1267–77.

[41] Commissions forcibly recruiting men for military service.
[42] Poor, inferior persons.

of all authority and provided for a democratic form of government, without king or house of lords, based on widespread manhood suffrage. Sovereign power was to be lodged in the representatives, except in regard to matters of religion, military conscription, punishment for participation in the recent conflict and equality before the law: these matters were reserved for the decision of the people.

Having, by our late labors and hazards, made it appear to the world at how high a rate we value our just freedom, and God having so far owned our cause as to deliver the enemies thereof into our hands, we do now hold ourselves bound, in mutual duty to each other, to take the best care we can for the future to avoid both the danger of returning into a slavish condition and the chargeable remedy of another war. For, as it cannot be imagined that so many of our countrymen would have opposed us in this quarrel if they had understood their own good, so may we hopefully promise to ourselves that, when our common rights and liberties shall be cleared, their endeavors will be disappointed that seek to make themselves our masters. Since, therefore, our former oppressions and not-yet-ended troubles have been occasioned, either by want of frequent national meetings in council, or by the undue and unequal constitution thereof, or by rendering those meetings ineffectual, we are fully agreed and resolved, God willing, to provide that hereafter our representatives be neither left to an uncertainty for times nor be unequally constituted, nor made useless for the ends for which they are intended. In order whereunto we declare and agree:

First, that, to prevent the many inconveniences apparently arising from the long continuance of the same persons in supreme authority, this present Parliament end and dissolve itself upon or before the last day of April 1649.

Secondly, that the people of England – being at this day very unequally distributed by counties, cities and boroughs for the election of their representatives – be indifferently proportioned; and, to this end, that the representative of the whole nation shall consist of 400 persons, or not above; and in each country and the places thereto subjoined there shall be chosen, to make up the said representative at all times, the several members here mentioned.[43]. . . . Provided that the first or second representative may, if they see cause, assign the remainder of the 400 representers, not hereby assigned, or so many of them as they shall see cause for, unto such counties as shall appear in this present distribution to have less than their due proportion. Provided also that, where any city or borough, to which one representer or more is assigned, shall be found in a due proportion not competent alone to elect a representer or the number of

[43] Here follow quotas assigned to English and Welsh constituencies.

representers assigned thereto, it is left to future representatives to assign such a number of parishes or villages near adjoining to such city or borough, to be joined therewith in the elections, as may make the same proportionable.

Thirdly, that the people do of course choose themselves a representative once in two years, and shall meet for that purpose upon the first Thursday in every second May, by eleven in the morning; and the representatives so chosen [are] to meet upon the second Thursday in the June following at the usual place in Westminster, or such other place as, by the foregoing representative or the council of state in the interval, shall be from time to time appointed or published to the people, at the least twenty days before the time of election; and [are] to continue their sessions there or elsewhere until the second Thursday in December following, unless they shall adjourn or dissolve themselves sooner, but not to continue longer. The election of the first representative [is] to be on the first Thursday in May 1649; and that and all future elections [are] to be according to the rules prescribed for the same purpose in this agreement, viz.: (1) That the electors in every division shall be natives or denizens of England; not persons receiving alms, but such as are assessed ordinarily towards the relief of the poor; not servants to, and receiving wages from, any particular person; and in all elections, except for the universities, they shall be men of twenty-one years of age or upwards and housekeepers, dwelling within the division for which the election is. . . . (4) That, to the end all officers of state may be certainly accountable and no factions made to maintain corrupt interests, no member of a council of state, nor any officer of any salary-forces in army or garrison, nor any treasurer or receiver of public money, shall, while such, be elected to be of a representative; and in case any such election shall be, the same to be void; and in case any lawyer shall be chosen into any representative or council of state, then he shall be incapable of practice as a lawyer during that trust. (5) For the more convenient election of representatives, each county wherein more than three representers are to be chosen, with the towns corporate and cities . . . within the compass thereof, to which no representers are herein assigned, shall be divided by a due proportion into so many and such parts as each part may elect two, and no part above three representers. . . .

Fourthly, that 150 members at least be always present in each sitting of the representative at the passing of any law or doing of any act whereby the people are to be bound; saving that the number of 60 may make a house for debates or resolutions that are preparatory thereunto.

Fifthly, that each representative shall, within twenty days after their first meeting, appoint a council of state for the managing of public affairs until the tenth day after the meeting of the next representative, unless that next representative think fit to put an end to that trust sooner. And the same council [is] to act and proceed therein according to such instructions and limitations as the representative shall give, and not otherwise.

Sixthly, that in each interval betwixt biennial representatives the council of state, in case of imminent danger or extreme necessity, may summon a representative to be forthwith chosen and to meet; so as the session thereof continue not above eighty days, and so as it dissolve at least fifty days before the appointed time for the next biennial representative. And upon the fiftieth day so preceding it shall dissolve of course, if not otherwise dissolved sooner.

* * *

Eighthly, that the representatives have . . . the supreme trust in order to the preservation and government of the whole; and that their power extend, without the consent or concurrence of any other person or persons, to the erecting and abolishing of courts of justice and public offices, and to the enacting, altering, repealing and declaring of laws, and the highest and final judgment concerning all natural or civil things, but not concerning things spiritual or evangelical. Provided that, even in things natural and civil, these six particulars next following are . . . excepted and reserved from our representatives, viz.: (1) We do not empower them to impress or constrain any person to serve in foreign war, either by sea or land, nor for any military service within the kingdom; save that they may take order for the forming, training and exercising of the people in a military way, to be in readiness for resisting of foreign invasions, suppressing of sudden insurrections, or for assisting in execution of the laws. And [they] may take order for the employing and conducting of them for those ends, provided that, even in such cases, none be compellable to go out of the county he lives in, if he procure another to serve in his room. (2) That, after the time herein limited for the commencement of the first representative, none of the people may be at any time questioned for anything said or done in relation to the late wars or public differences, otherwise than in execution or pursuance of the determinations of the present House of Commons, against such as have adhered to the King or his interest against the people; and saving that accountants for public moneys received shall remain accountable for the same. (3) That no securities given or to be given by the public faith of the nation, nor any engagements of the public faith for satisfaction of debts and damages, shall be made void or invalid by the next or any future representatives, except to such creditors as have or shall have justly forfeited the same; and saving that the next representative may confirm or make null, in part or in whole, all gifts of lands, moneys, offices or otherwise, made by the present Parliament to any member or attendant of either house. (4) That, in any laws hereafter to be made, no person, by virtue of any tenure, grant, charter, patent, degree or birth, shall be privileged from subjection thereto, or from being bound thereby, as well as others. (5) That the representative may not give judgment upon any man's person or estate, where no law hath before provided; save only in calling to account and punishing public officers for abusing or failing in their trust. (6) That no representative may in any wise render up or give or take away any of the foundations of common right, liberty and safety contained in this agreement, nor level men's estates, destroy property, or make all things common; and that, in all matters of such fundamental concernment, there shall be a liberty to particular members of the said representatives to enter their dissents from the major vote.

Ninthly, concerning religion, we agree as followeth: (1) It is intended that the Christian religion be held forth and recommended as the public profession in this nation; which we desire may, by the grace of God, be reformed to the greatest purity in doctrine, worship and discipline, according to the word of God. The instructing the people thereunto in a public way, so it be not compulsive, as also the maintaining of able teachers for that end and for the confutation or discovery of heresy, error and whatsoever is contrary to sound doctrine, is allowed to be provided for by our representatives; the maintenance of which teachers may be out of a public treasury, and, we desire, not by tithes – provided that popery or prelacy be not held forth as the public way or profession in this nation. (2) That to the public profession so held forth, none be compelled by penalties or otherwise, but only may be endeavored to be won by sound doctrine and the example of a good conversation. (3) That such as profess faith in God by Jesus Christ, however differing in judgment from the doctrine, worship or discipline publicly held forth, as aforesaid, shall not be restrained from, but shall be protected in, the profession of their faith and exercise of religion according to their consciences in any place except such as shall be set apart for the public worship . . . so as they abuse not this liberty to the civil injury of others or to actual disturbance of the public peace on their parts. Nevertheless, it is not intended to be hereby provided that this liberty shall necessarily extend to popery or prelacy. (4) That all laws, ordinances, statutes, and clauses in any law, statute or ordinance to the contrary of the liberty herein provided for, in the two particulars next preceding concerning religion, be and are hereby repealed and made void.

Tenthly, it is agreed that whosoever shall by force of arms resist the orders of the next or any future representative – except in case where such representative shall evidently render up, or give, or take away the foundations of common right, liberty and safety contained in this agreement – he shall forthwith . . . lose the benefit and protection of the laws and shall be punishable with death as an enemy and traitor to the nation. Of the things expressed in this agreement – the certain ending of this Parliament . . . ; the equal or proportionable distribution of the number of the representers to be elected . . . ; the certainty of the people's meeting to elect for representatives biennial, and their freedom in elections, with the certainty of meeting, sitting and ending of representatives so elected . . . ; as also the qualifications of persons to elect or be elected . . . ; also the certainty of a number for passing a law or preparatory debates . . . ; the matter of the fifth article, concerning the council of state; and of the sixth, concerning the calling, sitting and ending of representatives extraordinary; also the power of representatives to be, as in the eighth article, and limited, as in the six reserves next following the same; likewise the second and third particulars under the ninth article concerning religion; and the whole matter of the tenth article – all these we do account and declare to be fundamental to our common right, liberty and safety; and therefore do both agree thereunto and resolve to maintain the same, as God shall enable us. The rest of the matters in this agreement we account to be useful and good for the public. And the particular circumstance of numbers, times and places expressed in the several articles we account not fundamental; but we find them necessary to be here determined for the making the agreement certain and practicable, and do hold these most convenient that are here set down, and therefore do positively agree thereunto.

By the appointment of his Excellency the Lord General and his general council of officers.

179. Charles I on Trial, 1649

While democratic theories were gaining adherents, Charles I threw in his lot with the Scots, and the civil conflict briefly flamed up again. With the Scots, the extension of their Presbyterian system was the principal aim. Charles was primarily interested in the restoration of his authority, but was willing to agree that Presbyterianism should be established in England for a three-year trial period. But the Scots and English royalists were defeated, and the radical element took the offensive, bringing the King to trial before a special court. Since he refused to recognize its authority, he was not permitted to plead his case. But he issued a statement explaining his attitude toward the court, arguing that popular liberties were bound up with his own, and

---

SOURCE: John Rushworth, *Historical Collections* (London: J. A. for R. Boulter, 1680–1722), *VII*, 1403–04.

that for him to acknowledge an illegal jurisdiction would be to betray those liberties.

Having already made my protestations, not only against the illegality of this pretended court, but also that no earthly power can justly call me, who am your king, in question as a delinquent, I would not any more open my mouth upon his occasion more than to refer myself to what I have spoken, were I in this case alone concerned; but the duty I owe to God in the preservation of the true liberty of my people will not suffer me at this time to be silent. For how can any freeborn subject of England call life or anything he possesseth his own if power without right daily make new and abrogate the old fundamental laws of the land – which I now take to be the present case? . . .

There is no proceeding just against any man but what is warranted either by God's laws or the municipal laws[44] of the country where he lives. Now I am most confident this day's proceeding cannot be warranted by God's laws; for, on the contrary, the authority of obedience unto kings is clearly warranted and strictly commanded in both the Old and New Testament, which, if denied, I am ready instantly to prove. . . . Then, for the law of this land, I am no less confident that no learned lawyer will affirm that an impeachment can lie against the King, they all going in his name. And one of their maxims is that the King can do no wrong. Besides, the law upon which you ground your proceedings must either be old or new: if old, show it; if new, tell what authority, warranted by the fundamental laws of the land, hath made it, and when.

How the House of Commons can erect a court of judicature, which was never one itself . . . , I leave to God and the world to judge. And it were full as strange that they should pretend to make laws without King or Lords' House, to any that have heard speak of the laws of England. And admitting, but not granting, that the people of England's commission could grant your pretended power, I see nothing you can show for that; for certainly you never asked the question of the tenth man in the kingdom, and in this way you manifestly wrong even the poorest ploughman, if you demand not his free consent. . . .

Thus you see that I speak not for my own right alone . . . , but also for the true liberty of all my subjects, which consists, not in the power of government, but in living under such laws, such a government, as may give themselves the best assurance of their lives and property of their goods . . . Besides all this, the peace of the kingdom is not the least in my thoughts. And what hope of settlement is there so long as power reigns without rule or law, changing the whole frame of that government

[44] Laws pertaining to the internal affairs of the nation.

under which this kingdom hath flourished for many hundred years? . . . And believe it, the commons of England will not thank you for this change; for they will remember how happy they have been of late years under the reigns of Queen Elizabeth, the King my father, and myself, until the beginnings of these unhappy troubles, and will have cause to doubt that they shall never be so happy under any new. And by this time it will be too sensibly evident that the arms I took up were only to defend the fundamental laws of this kingdom against those who have supposed my power hath totally changed the ancient government.

Thus, having showed you briefly the reasons why I cannot submit to your pretended authority without violating the trust which I have from God for the welfare and liberty of my people, I expect from you either clear reasons to convince my judgment . . . , or that you will withdraw your proceedings.

This I intended to speak in Westminster Hall on Monday, January 22, but against reason was hindered to show my reasons.

180. Charles I's Death Sentence

The High Court of Justice found Charles I guilty of high treason and other crimes, and sentenced him to death. That a king could be regarded as guilty of treason ran counter to all traditions of English law, but the composition of the court, reflecting the prevailing military ascendancy, as well as its procedure, made such a verdict a foregone conclusion.

Whereas the Commons of England assembled in Parliament have by their late Act . . . authorized and constituted us an high court of justice for the trying and judging of . . . Charles Stuart for the crimes and treasons in the said Act mentioned; by virtue whereof the said Charles Stuart hath been three several times convented before this high court, where the first day . . . a charge of high treason and other crimes was, in the behalf of the people of England, exhibited against him and read openly unto him, wherein he was charged that he . . . , being admitted King of England and therein trusted with a limited power to govern by and according to the law of the land and not otherwise; . . . nevertheless, out of a wicked design to erect and uphold in himself an unlimited and tyrannical power to rule according to his will and to overthrow the rights and liberties of the people . . . , he . . . hath traitorously and maliciously levied war against the present Parliament and people therein represented . . . ; and that, by

---

SOURCE: John Rushworth, *Historical Collections* (London: J. A. for R. Boulter, 1680–1722), *VII*, 1418–19.

the said cruel and unnatural war so levied, continued and renewed, much innocent blood of the free people of this nation hath been spilt, many families undone, the public treasure wasted, trade obstructed and miserably decayed, vast expense and damage to the nation incurred, and many parts of the land spoiled, some of them even to desolation. . . . Whereupon the proceedings and judgments of this court were prayed against him as a tyrant, traitor and murderer and public enemy to the commonwealth, as by the said charge more fully appeareth. To which charge he . . . was required to give his answer, but he refused to do so; . . . upon which his several defaults this court might justly have proceeded to judgment against him, both for his contumacy and the matters of the charge . . .

Yet nevertheless this court, for its own clearer information and further satisfaction, have thought fit to examine witnesses upon oath and take notice of other evidences touching the matters contained in the said charge, which accordingly they have done.

Now, therefore, . . . this court is in judgment and conscience satisfied that he, the said Charles Stuart, is guilty of levying war against the said Parliament and people, and maintaining and continuing the same, for which in the said charge he stands accused; and, by the general course of his government, counsels and practices, . . . this court is fully satisfied in their judgments and consciences that he has been and is guilty of the wicked design and endeavors in the said charge set forth. . . . For all which treasons and crimes this court doth adjudge that he, the said Charles Stuart, as a tyrant, traitor, murderer and public enemy to the good people of this nation, shall be put to death by severing of his head from his body.

SECTION F

The Commonwealth and Protectorate

181. The Abolition of the Monarchy, 1649

Immediately after the execution of Charles, the Commons passed the following act abolishing the kingship. The House of Lords, which had refused to sanction the High Court of Justice, was also abolished, the

---

SOURCE: C. H. Firth and R. S. Rait (eds.), *Acts and Ordinances of the Interregnum* (London: Her Majesty's Stationery Office, 1911), *II*, 19–20.

Commons claiming, in resolutions early in January, that they, "being chosen by and representing the people, have the supreme power in this nation."[45] Another act formally designated the new regime as a "commonwealth and free state."

Whereas it is and hath been found by experience that the office of a king in this nation and Ireland . . . is unnecessary, burdensome, and dangerous to the liberty, safety and public interest of the people; and that for the most part use hath been made of the regal power and prerogative to oppress and impoverish and enslave the subject; and that usually and naturally any one person in such power makes it his interest to encroach upon the just freedom and liberty of the people and to promote the setting up of their own will and power above the laws, that so they might enslave these kingdoms to their own lust: be it therefore enacted and ordained by this present Parliament . . . that the office of a king in this nation shall not henceforth reside in or be exercised by any one single person. . . .

And whereas, by the abolition of the kingly office provided for in this act, a most happy way is made for this nation, if God see it good, to return to its just and ancient right of being governed by its own representatives or national meetings in council, from time to time chosen and entrusted for that purpose by the people: it is therefore resolved and declared by the Commons assembled in Parliament that they will put a period to the sitting of this present Parliament and dissolve the same so soon as may possibly stand with the safety of the people that hath betrusted them, and with what is absolutely necessary for the preserving and upholding the government now settled in the way of a commonwealth; and that they will carefully provide for the certain choosing, meeting, and sitting of the next and future representatives, with such other circumstances of freedom in choice and equality in distribution of members to be elected thereunto as shall most conduce to the lasting freedom and good of this commonwealth.

And it is hereby further enacted and declared, notwithstanding anything contained in this act, [that] no person or persons of what condition and quality soever within the Commonwealth of England and Ireland, Dominion of Wales, the islands of Guernsey and Jersey, the town of Berwick-upon-Tweed, shall be discharged from the obedience and subjection which he and they owe to the government of this nation, as it is now declared; but all and every of them shall in all things render and perform the same, as of right is due unto the supreme authority hereby declared to reside in this and the successive representatives of the people of this nation, and in them only.

[45] *Commons Journals*, VI, 111.

182. The Instrument of Government, 1653

The governmental arrangements of 1649 were altered in 1653 to permit the exercise of administrative authority by a single person, called the Lord Protector, advised by a council of state. A single-chamber parliament continued to serve the legislative function until 1657, when a modified House of Lords was restored. The document embodying the new constitution, and naming Oliver Cromwell as Protector, was called the Instrument of Government.

The government of the Commonwealth of England, Scotland and Ireland, and the dominions thereunto belonging.

1. That the supreme legislative authority of the Commonwealth of England, Scotland and Ireland, and the dominions thereunto belonging, shall be and reside in one person and the people assembled in Parliament; the style of which person shall be the Lord Protector of the Commonwealth of England, Scotland and Ireland.

2. That the exercise of the chief magistracy and the administration of the government over the said countries and dominions, and the people thereof, shall be in the Lord Protector, assisted with a council, the number whereof shall not exceed twenty-one nor be less than thirteen.

3. That all writs, process, commissions, patents, grants and other things, which now run in the name and style of the Keepers of the Liberty of England by Authority of Parliament, shall run in the name and style of the Lord Protector, from whom, for the future, shall be derived all magistracy and honors in these three nations. And [he shall] have the power of pardons, except in case of murders and treason, and benefit of all forfeitures for the public use; and shall govern the said countries and dominions in all things by the advice of the council, and according to these presents and the laws.

4. That the Lord Protector, the Parliament sitting, shall dispose and order the militia and forces, both by sea and land, for the peace and good of the three nations by consent of Parliament; and that the Lord Protector, with the advice and consent of the major part of the Council, shall dispose and order the militia for the ends aforesaid in the intervals of Parliament.

5. That the Lord Protector, by the advice aforesaid, shall direct in all things concerning the keeping and holding of a good correspondence with foreign kings, princes and states; and also with the consent of the major part of the Council, have the power of war and peace.

6. That the laws shall not be altered, suspended, abrogated or repealed,

---

SOURCE: C. H. Firth and R. S. Rait (eds.), *Acts and Ordinances of the Interregnum* (London: Her Majesty's Stationery Office, 1911), *II*, 813–22.

nor any new law made, nor any tax, charge or imposition laid upon the people, but by common consent in Parliament, save only as is expressed in the thirtieth article.

7. That there shall be a Parliament summoned to meet at Westminster upon the 3d day of September 1654, and that successively a Parliament shall be summoned once in every third year . . .

8. That neither the Parliament to be next summoned nor any successive Parliaments shall, during the time of five months to be accounted from the day of their first meeting, be adjourned, prorogued or dissolved without their own consent.

9. That as well the next as all other successive Parliaments shall be summoned and elected in manner hereafter expressed: that is to say, the persons to be chosen within England, Wales, the isles of Jersey, Guernsey and the town of Berwick-upon-Tweed to sit and serve in Parliament shall be, and not exceed, the number of 400; the persons to be chosen within Scotland . . . the number of 30; and the persons to be chosen . . . for Ireland . . . the number of 30. . . .

14. That all and every person and persons who have aided, advised, assisted or abetted in any war against Parliament since the first day of January 1641 – unless they have been since in the service of the Parliament and given signal testimony of their good affection thereunto – shall be disabled and incapable to be elected or to give any vote in the election of any members to serve in the next Parliament or in the three succeeding triennial Parliaments.

15. That all such who have advised, assisted or abetted the rebellion of Ireland shall be disabled and incapable forever to be elected or give any vote in the election of any member to serve in Parliament; as also all such who do or shall profess the Roman Catholic religion. . . .

17. That the persons who shall be elected to serve in Parliament shall be such . . . as are persons of known integrity, fearing God, and of good conversation, and being of the age of twenty-one years.

18. That all and every person and persons seised or possessed to his own use of any estate, real or personal, to the value of £200, and not within the aforesaid exceptions, shall be capable to elect members to serve in Parliament for counties. . . .

24. That all bills agreed unto by the Parliament shall be presented to the Lord Protector for his consent; and in case he shall not give his consent thereto within twenty days after they shall be presented to him, or give satisfaction to the Parliament within the time limited, that then, upon declaration of the Parliament that the Lord Protector hath not consented nor given satisfaction, such bills shall pass into and become laws, . . . provided such bills contain nothing in them contrary to the matters contained in these presents.

25. That Henry Lawrence [with 14 others] . . . , or any seven of them, shall be a council for the purposes expressed in this writing. And upon the death or other removal of any of them, the Parliament shall nominate six persons of ability, integrity, and fearing God, for every one that is dead or removed, out of which the major part of the council shall elect two, and present them to the Lord Protector, of whom he shall elect one. . . .

26. That the Lord Protector and the major part of the Council aforesaid may, at any time before the meeting of the next Parliament, add to the Council such persons as they shall think fit, provided the number of the Council be not made thereby to exceed twenty-one . . .

27. That a constant yearly revenue shall be raised, settled and established for maintaining of 10,000 horse and dragoons[46] and 20,000 foot in England, Scotland and Ireland, for the defense and security thereof; and also for a convenient number of ships for guarding of the seas; besides £200,000 per annum for defraying the other necessary charges of administration of justice and other expenses of the government – which revenue shall be raised by the customs, and such other ways and means as shall be agreed upon by the Lord Protector and the Council, and shall not be taken away or diminished, nor the way agreed upon for raising the same altered, but by the consent of the Lord Protector and the Parliament.

32. That the office of Lord Protector over these nations shall be elective and not hereditary and, upon the death of the Lord Protector, another fit person shall be forthwith elected to succeed him in the government; which election shall be by the Council. . . .

33. That Oliver Cromwell, Captain-General of the forces of England, Scotland and Ireland, shall be and is hereby declared to be Lord Protector of the Commonwealth of England, Scotland and Ireland, and the dominions thereto belonging, for his life.

34. That the chancellor, keeper or commissioners of the great seal, the treasurer, admiral, chief governors of Ireland and Scotland, and the chief justices of both the benches, shall be chosen by the approbation of Parliament; and in the intervals of Parliament, by the approbation of the major part of the council, to be afterwards approved by the Parliament.

35. That the Christian religion, as contained in the Scriptures, be held forth and recommended as the public profession of these nations; and that, as soon as may be, a provision less subject to scruple and contention, and more certain than the present, be made for the encouragement and maintenance of able and painful teachers, for instructing the people and for discovery and confutation of error, heresy and whatever is contrary to sound doctrine; and that, until such provision be made, the present maintenance shall not be taken away nor impeached.

[46] Cavalrymen and mounted infantrymen.

36. That to the public profession held forth none shall be compelled by penalties or otherwise, but that endeavors be used to win them by sound doctrine and the example of a good conversation.

37. That such as profess faith in God by Jesus Christ, though differing in judgment from the doctrine, worship or discipline publicly held forth, shall not be restrained from, but shall be protected in, the profession of the faith and exercise of their religion; so as they abuse not this liberty to the civil injury of others and to the actual disturbance of the public peace on their parts; provided this liberty be not extended to popery nor prelacy, nor to such as, under the profession of Christ, hold forth and practice licentiousness. . . .

41. That every successive Lord Protector over these nations shall take and subscribe a solemn oath, in the presence of the Council and such others as they shall call to them, that he will seek the peace, quiet and welfare of these nations, cause law and justice to be equally administered, and that he will not violate or infringe the matters and things contained in this writing, and in all other things will, to his power and to the best of his understanding, govern these nations according to the laws, statutes and customs thereof. . . .

183. Cromwell's "Fundamentals," 1654

Cromwell's philosophy of government laid stress on the importance of joint control by the administrative head of the state and the legislature, and on a large measure of religious toleration. The Instrument of Government gives evidence of this, as does the following speech made by Cromwell to the first Protectorate Parliament in 1654, in which he specifically refers to a single chief executive, parliaments without permanent tenure, liberty of conscience, and joint control of the militia by Protector and Parliament, as inalterable bases of government.

It is true, there are some things in the Establishment that are fundamental, and some things are not so, but are circumstantial. Of such,[47] no question but I shall easily agree to vary, or leave out, as I shall be convinced by reason. Some things are fundamental, about which I shall deal plainly with you; they may not be parted with, but will (I trust) be delivered over to posterity, as being the fruits of our blood and travel.[48]

SOURCE: Wilbur C. Abbott, *Writings and Speeches of Oliver Cromwell* (Cambridge: Harvard University Press, 1937–47), *III*, 458–60.

---

[47] I.e., of circumstantial matters.
[48] Travail.

The government by a single person and a Parliament is a fundamental; it is the *esse*,[49] it is constitutive. And for the person, though I may seem to plead for myself, yet I do not, no, nor can any reasonable man say it . . . In every government there must be somewhat fundamental, somewhat like a Magna Carta, that should be standing and be unalterable. Where there is a stipulation on one part, and that fully accepted, as appears by what hath been said, surely a return ought to be: else what does that stipulation signify? If I have upon the terms aforesaid undertaken this great trust, and exercised it, and by it called you, surely it ought to be owned.[50]

That Parliaments should not make themselves perpetual is a fundamental. Of what assurance is a law to prevent so great an evil, if it lie in one or the same legislator to unlaw it again?[51] Is this like to be lasting? It will be like a rope of sand; it will give no security, for the same men may unbuild what they have built.

Is not liberty of conscience in religion a fundamental? So long as there is liberty of conscience for the supreme magistrate to exercise his conscience in erecting what form of church-government he is satisfied he should set up, why should not he give it to others? Liberty of conscience is a natural right; and he that would have it ought to give it, having liberty to settle what he likes for the public.

Indeed, that hath been one of the vanities of our contests. Every sect saith, Oh! Give me liberty. But give him it, and to his power he will not yield it to anybody else. . . . And I may say it to you, I can say it: All the money of this nation would not have tempted men to fight upon such an account as they have engaged, if they had not had hopes of liberty, better than they had from Episcopacy, or than would have been afforded them from a Scottish Presbytery, or an English either, if it had made such steps or been as sharp and rigid as it threatened when it was first set up. This I say is a fundamental. It ought to be so: it is for us, and the generations to come. And if there be an absoluteness in the imposer, without fitting allowances and exceptions from the rule, we shall have our people driven into wildernesses, as they were when those poor and afflicted people, that forsook their estates and inheritances here, where they lived plentifully and comfortably, for the enjoyment of their liberty, and were necessitated to go into a vast howling wilderness in New England, where they have for liberty sake stript themselves of all their comfort and the full enjoyment they had, embracing rather loss of friends and want, than to be so ensnared and in bondage.

[49] Being.
[50] In other words, Cromwell, having undertaken the headship of the state under certain conditions, should be able to count on the permanence of such an arrangement.
[51] Revoke it.

Another, which I had forgotten, is the militia; that's judged a fundamental, if anything be so. That it should be well and equally placed, is very necessary. For put the absolute power of the militia into one without a check, what doth it? . . . What signifies a provision against perpetuating of Parliaments, if this be solely in them? Whether, without a check, the Parliament have not liberty to alter the frame of government to aristocracy, to democracy, to anarchy, to anything, if this be fully in them, yea, into all confusion, and that without remedy? And if this one thing be placed in one; that one, be it Parliament, be it supreme governor, they or he hath power to make what they please of all the rest.

Therefore, if you would have a balance at all, and that some fundamentals must stand which may be worthy to be delivered over to posterity, truly I think it is not unreasonably urged, that the militia should be disposed, as it is laid down in the [Act of] Government and that it should be so equally placed that one person neither in Parliament nor out of Parliament, should have the power of ordering it. . . .

SECTION G

Opposition and Unrest, 1649–1660

184. A Leveller Attack on the Commonwealth, 1649

The Commonwealth and Protectorate were opposed not only by royalists, but also by others who found that the work of Cromwell and his aides fell short of what they desired in the way of political and social reconstruction. Early in 1649 John Lilburne, leader of the Levellers, was already attacking the Commonwealth as an illiberal regime. The following passages are from his pamphlet, *England's New Chains.*

Where is that good, or where is that liberty so much pretended, so dearly purchased? If we look upon what this House hath done since it hath voted itself the supreme authority, and disburthened themselves of the power of the Lords. First, we find a high court of justice erected,[52]

SOURCE: John Lilburne, *England's New Chains Discovered* (London: 1649), in Margaret James and Maureen Weinstock, *England during the Interregnum, 1642–1660* (London: Longmans, Green, 1935), 152–53.

52 A special court, set up by Parliament in February 1649, to try five notorious royalists.

or trial of criminal causes, whereby that great and strong hold of our preservation, the way of trial by twelve sworn men of the neighborhood, is infringed, all liberty of exception against the triers is overruled by a court consisting of persons picked and chosen in an unusual way; the practice whereof we cannot allow of, though against open and notorious enemies, as well because we know it to be an usual policy to introduce by such means all usurpations, first against adversaries in hope of easier admission, as also for that the same being so admitted may at pleasure be exercised against any person or persons whatsoever. This is the first part of our new liberty. The next is the censuring of a member of this House for declaring his judgment in a point of religion. . . . Besides there is the act for pressing of seamen, directly contrary to the agreement of the officers. Then the stopping of our mouths from printing is carefully provided for. . . . Then, whereas it was expected that the Chancery and courts of justice in Westminster and the judges and officers thereof should have been surveyed and for the present regulated . . . instead thereof the old and advanced fees are continued and new thousand pounds annual stipends allotted; when in the corruptest times the ordinary fees were thought a great and a sore burden. . . . What now is become of that liberty that no man's person shall be attacked or imprisoned, or otherwise disseised of his freehold or free customs, but by lawful judgment of his equals?

185. The Dissatisfaction of Fifth Monarchy Men, 1655

Once Cromwell became Protector, he was assailed by the Fifth Monarchy Men, a sect of religious enthusiasts who envisioned the establishment of a theocratic republic and the speedy advent of the millenium. They objected not only to the monarchical tendencies in the Protectorate, but also to Cromwell's maintenance of a national church, preaching revolt against the regime and even attempting a rising. These passages are taken from one of their tracts, *A Word for God* (1655).

We do also believe in our heart that (though the worst things are not without God's permission and providence), yet that this government is not of God's approbation or taken up by His counsel or according to His word; and therefore we do utterly disclaim having any hand or heart in it, and for the contrivers and undertakers thereof, we suspect and judge them to be great transgressors therein; and so much the more because

---

SOURCE: Thomas Birch (ed.), *A Collection of the State Papers of John Thurloe* (London: Executor of Fletcher Gyles, 1742), *IV*, 382.

they are professors of religion, and declarers, engagers and fighters against the very things they now practise, and it is most evident to us that they thereby build again what before they did destroy; and in so doing they render themselves and the cause, religion, name and people of God abominable to the heathens, papists and profane enemies, which is a grief to our souls to consider. We do also detest the practices of those men in imprisoning the saints of God for their consciences and testimony, and just men, who stand for moral and just principles and the freedom of the nation and people, and their breaking off parliaments to effect their own design. We also from our souls witness against their new modelling of ministers (as anti-Christian) and keeping up parishes and tithes (as popish innovations). . . .

186. Winstanley's Agrarian Communism, 1656

A few men were not satisfied with political and religious reform, but sought changes which would better the economic condition of the rank and file. One small group, known as the "Diggers," preached a sort of agrarian communism, and actually attempted to appropriate and cultivate unenclosed lands. One of their leaders was Gerard Winstanley. In his *New Law of Righteousness*, he wrote:

Divide England into three parts, scarce one part is manured; so that here is land enough to maintain all her children, and many die for want or live under a heavy burden of poverty all their days. And this misery the poor people have brought upon themselves, by lifting up particular interest by their labors.

There are yet three doors of hope for England to escape destroying plagues:

First, let everyone leave off running after others for knowledge and comfort, and wait upon the spirit Reason, till he break forth out of the clouds of your heart and manifest himself within you. This is to cast off the shadow of learning, and to reject covetous, subtle, proud flesh that deceives all the world by their hearsay and traditional preaching of words, letters and syllables, without the spirit; and to make choice of the Lord, the true teacher of everyone in their own inward experience . . .

Secondly, let everyone open his bags and barns, that all may feed upon the crops of the earth, that the burden of poverty may be removed. Leave off this buying and selling of land, or of the fruits of the earth; and as it was in the light of reason first made, so let it be in action,

---

SOURCE: George H. Sabine (ed.), "The New Law of Righteousness," *Works of Gerrard Winstanley* (Ithaca: Cornell University Press, 1941), 200–01.

amongst all a common treasury; none enclosing or hedging in any part of earth, saying, this is mine – which is rebellion and high treason against the King of Righteousness. And let this word of the Lord be acted amongst all: work together, eat bread together. . . .

Thirdly, leave off dominion and lordship one over another, for the whole bulk of mankind are but one living earth. Leave off imprisoning, whipping and killing, which are but the actings of the curse. And let those that hitherto have had no land and have been forced to rob and steal through poverty, hereafter let them quietly enjoy land to work upon, that everyone may enjoy the benefit of his creation, and eat his own bread with the sweat of his own brows. For surely this particular propriety of mine and thine hath brought in all misery upon people. For first, it hath occasioned people to steal one from another. Secondly, it hath made laws to hang those that did steal. It tempts people to do an evil action, and then kills them for doing of it: let all judge if this be not a great devil.

SECTION H

The Restoration Settlement

187. The Declaration of Breda, 1660

Republican government did not long outlive Oliver Cromwell, who died in September, 1658. Thereafter a chaotic situation developed, in which ambitious army officers and the restored Rump Parliament jostled for position. This led to the restoration of the monarchy, as the only apparent means of restoring stability and avoiding military dictatorship. The way was paved by the Declaration of Breda, which the exiled Charles II issued from Holland as proof of his intentions regarding a general pardon, the religious question, rights in land, and the payment of the army. Early in May Charles was proclaimed King.

Charles, by the grace of God King of England, Scotland, France and Ireland, Defender of the Faith, etc., to all our loving subjects, . . . greeting. If the general distraction and confusion which is spread over the whole

---

SOURCE: *Journals of the House of Lords* (London: Her Majesty's Stationery Office), XI, 7–8.

kingdom doth not awaken all men to a desire and longing that those wounds which have so many years together been kept bleeding may be bound up, all we can say will be to no purpose. However, after this long silence, we have thought it our duty to declare how much we desire to contribute thereunto; and that, as we can never give over the hope in good time to obtain possession of that right which God and nature have made our due, so we do make it our daily suit to the Divine Providence that He will, in compassion to us and our subjects after so long misery and sufferings, remit and put us into a quiet and peaceable possession of that our right, with as little blood and damage to our people as is possible. Nor do we desire more to enjoy what is ours than that all our subjects may enjoy what by law is theirs by a full and entire administration of justice throughout the land, and by extending our mercy where it is wanted and deserved.

And to the end that the fear of punishment may not engage any conscious to themselves of what is past to a perseverance in guilt for the future, by opposing the quiet and happiness of their country in the restoration both of king, peers, and people to their just, ancient, and fundamental rights, we do by these presents declare that we do grant a free and general pardon to all our subjects, of what degree or quality soever, who, within forty days after the publishing hereof, shall lay hold upon this our grace and favor and shall, by any public act, declare their doing so and that they return to the loyalty and obedience of good subjects – excepting only such persons as shall hereafter be excepted by Parliament. Those only excepted, let all our subjects, how faulty soever, rely upon the word of a king, solemnly given by this present declaration, that no crime whatsoever, committed against us or our royal father before the publication of this, shall ever rise in judgment or be brought in question against any of them to the least endamagement of them, either in their lives, liberties, or estates, or – as far forth as lies in our power – so much as to the prejudice of their reputations by any reproach or term of distinction from the rest of our best subjects; we desiring and ordaining that henceforth all notes of discord, separation, and difference of parties be utterly abolished among all our subjects, whom we invite and conjure to a perfect union among themselves under our protection, for the resettlement of our just rights and theirs in a free Parliament, by which, upon the word of a king, we will be advised.

And because the passion and uncharitableness of the times have produced several opinions in religion, by which men are engaged in parties and animosities against each other – which, when they shall hereafter unite in a freedom of conversation, will be composed or better understood – we do declare a liberty to tender consciences and that no man shall be disquieted or called in question for differences of opinion in matter of religion, which do not disturb the peace of the kingdom; and that we shall be ready to consent to such an act of parliament as, upon mature deliberation, shall be offered to us for the full granting that indulgence.

And because, in the continued distractions of so many years and so many and great revolutions, many grants and purchases of estates have been made to and by many officers, soldiers, and others, who are now possessed of the same and who may be liable to actions at law upon several titles, we are likewise willing that all such differences and all things relating to such grants, sales, and purchases shall be determined in Parliament, which can best provide for the just satisfaction of all men who are concerned.

And we do further declare that we will be ready to consent to any act or acts of Parliament to the purposes aforesaid and for the full satisfaction of all arrears due to the officers and soldiers of the army under the command of General Monk;[53] and that they shall be received into our service upon as good pay and conditions as they now enjoy.

188. The Corporation Act, 1661

As suggested by the Declaration of Breda, the Convention Parliament of 1660 arranged an amnesty, worked out a land settlement and provided for the payment and dispersal of the Cromwellian army. But the religious settlement was left for the following Parliament, which was highly royalist and Anglican in sentiment. It was this "Cavalier" Parliament that restored the church on its Elizabethan foundations, took steps against nonconformist worship, and excluded those who would not take communion in the national church from municipal office. The third of these policies was implemented by the Corporation Act, which remained unrepealed until the nineteenth century.

Whereas questions are likely to arise concerning the validity of elections of magistrates and other officers and members in corporations, as well in respect of removing some as placing others during the late troubles contrary to the true intent and meaning of their charters and liberties;

SOURCE: *Statutes of the Realm* (London: Her Majesty's Stationery Office, 1810–28), V, 321–23.

---

[53] George Monk, whose military and political influence paved the way for the restoration of the monarchy.

and to the end that the succession in such corporations may be most probably perpetuated in the hands of persons well affected to his Majesty and the established government, it being too well known that notwithstanding all his Majesty's endeavors and unparalleled indulgence in pardoning all that is past . . . many evil spirits are still working . . . :

Be it enacted . . . that no charter of any corporation cities, towns, boroughs, Cinque Ports and their members, and other port towns in England and Wales . . . shall at any time hereafter be avoided[54] for or by reason of any act or thing done or omitted to be done before the first day of this present Parliament.

And be it further enacted by the authority aforesaid that all persons who, upon the four-and-twentieth day of December 1661, shall be mayors, aldermen, recorders,[55] bailiffs, town clerks, common councilmen and other persons then bearing any office or offices of magistracy or places or trusts or other employment relating to or concerning the government of the said respective cities, corporations and boroughs . . . shall . . . take the oaths of allegiance and supremacy and this oath following: "I, A. B., do declare and believe that it is not lawful upon any pretense whatsoever to take arms against the King, and that I do abhor that traitorous position of taking arms by his authority against his person or against those that are commissioned by him; so help me God." And also at the same time [he] shall publicly subscribe . . . this following declaration "I, A. B., do declare that I hold that there lies no obligation upon me or any other person from the oath commonly called the Solemn League and Covenant,[56] and that the same was in itself an unlawful oath and imposed upon the subjects of this realm against the known laws and liberties of the kingdom.". . .

Provided also, and be it enacted by the authority aforesaid, that from and after the expiration of the said commissions, no person or persons shall forever hereafter be placed, elected or chosen in or to any the offices or places aforesaid that shall not have within one year next before such election or choice taken the sacrament of the Lord's Supper according to the rites of the Church of England; and that every such person and persons so placed, elected or chosen shall likewise take the aforesaid three oaths and subscribe the said declaration at the same time when the oath for the due execution of the said places and offices respectively shall be administered. And in default hereof, every such placing, election and choice is hereby enacted and declared to be void. . . .

54 Made void.
55 The chief judicial officer of certain municipalities was known as the recorder.
56 See p. 262.

189. The Act of Uniformity, 1662

The restoration of the church rested essentially on a new Act of Uniformity.[57] This required, among other things, that clergymen and members of the teaching profession assent to doctrines set forth in the Book of Common Prayer and declare against resistance to the crown, as illustrated below. Many clergymen refused to comply with these provisions, and hence lost their livings and became known as Nonconformists.

And . . . be it further enacted . . . that every parson, vicar or other minister whatsoever, who now hath or enjoyeth any ecclesiastical benefice or promotion within the realm of England or places aforesaid, shall, in the church, chapel or place of public worship belonging to his said benefice or promotion, upon some Lord's day before the Feast of Saint Bartholomew[58] [1662] . . . openly and publicly before the congregation there assembled declare his unfeigned assent and consent to the use of all things in the said Book [of Common Prayer] contained and prescribed, in these words, and no other: "I, A. B., do here declare my unfeigned assent and consent to all and everything contained and prescribed in and by the book entitled *The Book of Common Prayer* . . ." And [it is enacted] that all and every such person who shall . . . neglect or refuse to do the same within the time aforesaid . . . shall *ipso facto* be deprived of all his spiritual promotions. . . .

And be it further enacted by the authority aforesaid that every dean, canon and prebendary of every cathedral or collegiate church, and all masters and other heads, fellows, chaplains and tutors of or in any college, hall, house of learning, or hospital, and every public professor and reader[59] in either of the universities, and in every college elsewhere, and every parson, vicar, curate, lecturer, and every other person in holy orders, and every schoolmaster keeping any public or private school, and every person instructing or teaching any youth in any house or private family as a tutor or schoolmaster, who upon the first day of May [1662] or at any time thereafter, shall be incumbent or have possession of any deanery, canonry [etc.] . . . shall . . . subscribe the declaration or acknowledgment following . . . : "I, A. B., do declare that it is not lawful, upon any pretense whatsoever, to take arms against the King, and that I do abhor

---

SOURCE: *Statutes of the Realm* (London: Her Majesty's Stationery Office, 1810–28), V, 365–66.

---

[57] For earlier Acts, see pp. 195, 199.
[58] August 24.
[59] Lecturer.

that traitorous position of taking arms by his authority against his person or against those that are commissionated[60] by him, and that I will conform to the liturgy of the Church of England as it is now by law established; and I do declare that I do hold there lies no obligation upon me or any other person, from the oath commonly called the Solemn League and Covenant, to endeavor any change or alteration of government either in church or state, and that the same was in itself an unlawful oath and imposed upon the subjects of this realm against the known laws and liberties of this kingdom.". . .

SECTION I

Religious Toleration and Persecution under Charles II

190. Some Consequences of the Conventicle Act, 1664

Although the various provisions of the Clarendon Code were not uniformly or regularly enforced, there is no doubt that they occasioned considerable distress. In his autobiography Richard Baxter, a noted contemporary Presbyterian clergyman, has this comment on conditions following the enactment of the Conventicle Act of 1664, by which nonconformist meetings were banned.

And now came in the people's trial, as well as the ministers'. While the danger and sufferings lay on the ministers alone, the people were very courageous and exhorted them to stand it out and preach till they went to prison. But when it came to be their own case, they were as venturous till they were once surprised and imprisoned; but then their judgments were much altered, and they that censured ministers before as cowardly, because they preached not publicly whatever followed, did now think that it was better to preach often in secret to a few than but once or twice in public to many; and that secrecy was no sin when it tended to the furtherance of the work of the Gospel and to the church's good. Espe-

SOURCE: J. M. Lloyd Thomas (ed.), *The Autobiography of Richard Baxter* (London: J. M. Dent and Sons, n.d.), 189.

60 Commissioned.

cially the rich were as cautelous[61] as the ministers. But yet their meetings were so ordinary and so well known that it greatly tended to the jailer's commodity. . . .

And here the fanatics called Quakers did greatly relieve the sober people for a time; for they were so resolute, and gloried in their constancy and sufferings, that they assembled openly (at the Bull and Mouth near Aldersgate) and were dragged away daily to the common jail; and yet desisted not, but the rest came the next day nevertheless. So that the jail at Newgate was filled with them. Abundance of them died in prison, and yet they continued their assemblies still. And the poor deluded souls would sometimes meet only to sit still in silence (when, as they said, the Spirit did not speak). And it was a great question whether this silence was a "religious exercise not allowed by the liturgy," etc. And once upon some such reasons as these, when they were tried at the sessions in order to a banishment, the jury acquitted them, but were grievously threatened for it. And after that another jury did acquit them, and some of them were fined and imprisoned for it. But thus the Quakers so employed Sir R. B. and the other searchers and prosecutors that they had the less leisure to look after the meetings of soberer men, which was much to their present ease. . . .

191. The Declaration of Indulgence, 1672

Charles II had his faults, but he was no persecutor. Moreover, political and diplomatic considerations induced him to attempt to remove the disabilities affecting non-Anglicans. In 1672 he issued the following Declaration of Indulgence, designed to suspend the operation of penal laws against Nonconformists and Roman Catholics.

Our care and endeavors for the preservation of the rights and interests of the church have been sufficiently manifested to the world by the whole course of our government since our happy restoration, and by the many and frequent ways of coercion that we have used for reducing all erring or dissenting persons and for composing the unhappy differences in matters of religion, which we found among our subjects upon our return. But, it being evident by the sad experience of twelve years that there is very little fruit of all those forcible courses, we think ourselves obliged

---

SOURCE: William Cobbett, *Parliamentary History of England* (London: R. Bagshaw [etc.], 1806–20), *IV*, 515–16.

---

[61] Cautious.

to make use of that supreme power in ecclesiastical matters which is not only inherent in us but hath been declared and recognized to be so by several statutes and acts of Parliament. And therefore we do now accordingly issue out this our royal declaration, as well for the quieting the minds of our good subjects in these points, for inviting strangers in this conjuncture to come and live under us, and for the better encouragement of all to a cheerful following of their trades and callings – from whence we hope, by the blessing of God, to have many good and happy advantages to our government – as also for preventing for the future the danger that might otherwise arise from private meetings and seditious conventicles.

And in the first place, we declare our express resolution, meaning and intention to be that the Church of England be preserved and remain entire in its doctrine, discipline and government, as now it stands established by law; and that this be taken to be, as it is, the basis, rule and standard of the general and public worship of God; and that the orthodox conformable clergy do receive and enjoy the revenues belonging thereunto; and that no person, though of different opinion or persuasion, shall be exempt from paying his tithes or other dues whatsoever. And further we declare that no person shall be capable of holding any benefice, living or ecclesiastical dignity or preferment of any kind in this kingdom of England, who is not exactly conformable.

We do in the next place declare our will and pleasure to be that the execution of all and all manner of penal laws in matters ecclesiastical against whatsoever sort of nonconformists or recusants be immediately suspended, and they are hereby suspended. And all judges of assize and jail-delivery, sheriffs, justices of the peace, mayors, bailiffs and other officers whatsoever, whether ecclesiastical or civil, are to take notice of it, and pay due obedience thereunto. And, that there may be no pretence for any of our subjects to continue their illegal meetings and conventicles, we do declare that we shall from time to time allow a sufficient number of places, as shall be desired, in all parts of this our kingdom, for the use of such as do not conform to the Church of England to meet and assemble in, in order to their public worship and devotion – which places shall be open and free to all persons. But to prevent such disorders and inconveniences as may happen by this our indulgence, if not duly regulated, and that they may be better protected by the civil magistrate, our express will and pleasure is that none of our subjects do presume to meet in any place until such place be allowed and the teacher of that congregation be approved by us. And lest any should apprehend that this restriction should make our said allowance and approbation difficult to be obtained, we do further declare that this our indulgence as to the allowance of public places of worship, and approbation of teachers, shall extend to all sorts of nonconformists and recusants, except the recusants[62] of the Roman Catholic religion, to whom we shall no ways allow in public places of worship, but only indulge them their share in the common exemption from the executing the penal laws and the exercise of their worship in their private houses only. And if, after this our clemency and indulgence, any of our subjects shall presume to abuse this liberty and shall preach seditiously, or to the derogation of the doctrine, discipline or government of the established church, or shall meet in places not allowed by us, we do hereby give them warning and declare we will proceed against them with all imaginable severity. And we will let them see we can be as severe to punish such offenders, when so justly provoked, as we are indulgent to truly tender consciences.

192. Reaction to the Declaration of Indulgence

The Declaration of Indulgence encountered pronounced opposition in Parliament, which would not concede that the King, alone, could legally suspend laws. It prevailed upon the King to withdraw it in return for a money grant and then proceeded to pass the Test Act (1673), excluding all who would not take communion in the Church of England from holding office under the crown. The following interchange of statements between Charles and the House of Commons is taken from the Commons Journals.

The King's Speech to Parliament – My Lords and Gentlemen:

Since you were last here I have been forced to a most important, necessary and expensive war, and I make no doubt but you will give me suitable and effectual assistance to go through with it. . . .

Some few days before I declared the war I put forth my declaration for indulgence to dissenters, and have hitherto found a good effect of it by securing peace at home when I had war abroad. There is one part of it that hath been subject to misconstruction, which is that concerning the papists, as if more liberty were granted them than to the other recusants, when it is plain there is less, for the others have public places allowed them, and I never intended that they should have any, but only have the freedom of their religion in their own houses, without any concern of others. And I could not grant them less than this when I had extended so much more grace to others, most of them having been loyal and in the

SOURCE: *Journals of the House of Commons* (London: Her Majesty's Stationery Office), *IX*, 246, 252.

[62] I.e., Catholics who refused to attend the services of the Established Church.

service of me and of the King my father. And in the whole course of this indulgence I do not intend that it shall any way prejudice the Church, but I will support its rights and it in its full power. Having said this, I shall take it very, very ill to receive contradiction in what I have done. And I will deal plainly with you: I am resolved to stick to my declaration.

There is one jealousy more that is maliciously spread abroad . . . and that is that the forces I have raised in this war were designed to control law and property. . . .

I will conclude with this assurance to you, that I will preserve the true reformed Protestant religion and the Church as it is now established in this kingdom, and that no man's property or liberty shall ever be invaded.

The Reply of the House of Commons – Most gracious Sovereign:

We, your Majesty's most loyal and faithful subjects, the Commons assembled in Parliament, do in the first place, as in all duty bound, return your Majesty our most humble and hearty thanks for the many gracious promises and assurances which your Majesty has several times during this present Parliament given to us, that your Majesty would secure and maintain unto us the true reformed Protestant religion, our liberties and properties, which most gracious assurances your Majesty hath out of your great goodness been pleased to renew unto us more particularly at the opening of this present session of Parliament.

And further, we crave leave humbly to represent that we have, with all duty and expedition, taken into our consideration several parts of your Majesty's last speech to us, and withal the declaration therein mentioned for indulgence to dissenters . . . and we find ourselves bound in duty to inform your Majesty that penal statutes in matters ecclesiastical cannot be suspended but by Act of Parliament.

We therefore, the knights, citizens and burgesses of your Majesty's House of Commons, do most humbly beseech your Majesty that the said laws may have their free course until it shall be otherwise provided for by Act of Parliament, and that your Majesty would graciously be pleased to give such directions herein that no apprehensions or jealousies may remain in the hearts of your Majesty's good and faithful subjects.

193. The Popish Plot, 1678

In 1678 anti-Catholic hysteria was fanned into flame by rumors spread by Titus Oates and others that a conspiracy was afoot to assassinate

SOURCES: Thora J. Stone, *England under the Restoration* (London: Longmans, Green, 1923), 44.

James Wellwood, *Memoirs of the Most Material Transactions in England* (London: T. Goodwin, 1700), 123–25.

Charles II, place the Catholic Duke of York upon the throne, and reintroduce the Catholic faith in England. Two contrasting impressions of the plot follow: one by the Duke of York himself, the other by the Whiggish James Wellwood.

As for news, this pretended plot is still under examination, and the judges are to give their opinion whether one witness in point of treason be sufficient to proceed criminously against anybody; and I do verily believe that when this affair is thoroughly examined it will be found nothing but malice against the poor Catholics in general, and myself in particular.

There is another thing happened, which is that a justice of the peace, Sir Edmund Berry Godfrey, was missing some days, suspected of several circumstances (very probable ones) to design the making himself away. Yesterday his body was found in a big place in the fields some two or three miles off, with his own sword through him. This makes a great noise and is laid upon the Catholics, also, but without any reason for it, for he was known to be far from being an enemy to them. . . . All these things happening together will cause, I am afraid, a great flame in this Parliament when they meet on Monday, for those disaffected to the Government will inflame things as much as they can.

That there was at that time a popish plot, and that there always has been one since the Reformation to support, if not restore, the Romish religion in England, scarce anybody calls in question. How far the near prospect of a popish successor[63] ripened the hopes and gave new vigor to the designs of that party, and what methods they were then upon to bring those designs about, Coleman's[64] letters alone, without any other concurring evidence, are more than sufficient to put the matter out of doubt. But what superstructures might have been afterwards built upon an unquestionable foundation, and how far some of the witnesses of that plot might come to darken truth by subsequent additions of their own, must be deferred till the Great Account, to be made before a higher tribunal . . . However, this is certain, the discovery of the Popish Plot had great and various effects upon the nation; and it's from this remarkable period of time we may justly reckon a new era in the English account.

In the first place, it awakened the nation out of a deep lethargy they had been in for nineteen years together, and alarmed them with fears and jealousies that have been found to our sad experience but too well grounded. In the next it gave rise to, at least settled that unhappy dis-

[63] James, Duke of York, who became James II in 1685.
[64] Edward Coleman, secretary to the Duchess of York, who carried on a correspondence with the French court.

tinction of Whig and Tory among the people of England, that has since occasioned so many mischiefs. And lastly, the discovery of the Popish Plot began that open struggle between King Charles and his people, that occasioned him not only to dissolve his first favorite Parliament, and the three others that succeeded, but likewise to call no more during the rest of his reign. All which made way for bringing in question the charters of London and other corporations, with a great many dismal effects that followed. . . .

SECTION J

Foreign Affairs under Charles II

194. The Declaration of War against the Dutch, 1665

Under Charles II, England fought two wars against the Dutch. Charles was for various reasons drawn into the orbit of Louis XIV, a development which (to the extent that it was publicly known) was regarded with increasing misgivings. The Anglo-Dutch conflict of 1664–1667 arose principally from commercial rivalry, as is seen in the declaration of war.

Whereas upon complaint of the several injuries, affronts and spoils done by the East and West India Companies and other the subjects of the United Provinces unto and upon the ships, goods and persons of our subjects, to their grievous damages and amounting to vast sums: instead of reparation and satisfaction, which hath been by us frequently demanded, we found that orders had been given to de Ruyter[65] not only to abandon the consortship[66] against the pirates of the Mediterranean Seas, to which the States General had invited us, but also to use all arts of depredation and hostility against our subjects in Africa.

We therefore gave orders for the detaining of the ships belonging to the States of the United Provinces, their subjects and inhabitants; yet

---

SOURCE: Thora J. Stone, *England under the Restoration* (London: Longmans, Green, 1923), 115.

---

[65] Michiel de Ruyter, Admiral-in-Chief of the Dutch fleet.
[66] Convoying arrangements.

notwithstanding we did not give any commission for letters of marque, nor were there any proceedings against the ships designed until we had a clear and undeniable evidence that de Ruyter had put the said orders in execution by seizing several of our subjects' ships and goods. . . .

We have thought fit, by and with the advice of our Privy Council, to declare, and do hereby declare to all the world, that the said States are the aggressors and that they ought in justice to be so looked upon by all men. So that as well our fleets and ships as also all other ships . . . that shall be commissionated by letters of marque from our dear brother the Duke of York, Lord High Admiral of England, shall and may lawfully fight with, subdue, seize and take all ships, vessels and goods belonging to the said States of the United Provinces, or any of their subjects or inhabitants.

195. The Secret Treaty of Dover, 1670

In 1670 Charles II and Louis XIV negotiated the secret Treaty of Dover. Charles agreed to assist France in another war against the Netherlands, in return for an annual grant from Louis and his promise to provide troops should Charles need them to reestablish Catholicism in England. Charles's financial reliance on Louis became a marked feature of his policy, and served to give him some independence in his relations with Parliament.

1. It is agreed, determined and concluded that there shall be forever a good, secure and firm peace, union, true fellowship, confederacy, friendship, alliance and good correspondence between the lord King of Great Britain, his heirs and successors on the one part, and the most Christian King Louis XIV on the other, and between all and every of their kingdoms, states and territories, as also between their subjects and vassals that they have or possess at present, or may have, hold and possess hereafter, as well by sea and fresh waters as by land. And as evidence that this peace shall remain inviolable, beyond the capacity of anything in the world to disturb it, there follow articles of so great confidence, and also so advantageous to the said lord Kings, that one will hardly find in any age more important provisions determined and concluded.

2. The lord King of Great Britain, being convinced of the truth of the Catholic religion, and resolved to declare it and reconcile himself with the Church of Rome as soon as the welfare of his kingdom will permit,

---

SOURCE: John Lingard, *History of England* (Dublin: James Duffy and Sons, 1874), *IX*, 251–54, in Paul L. Hughes and Robert F. Fries, *Crown and Parliament in Tudor-Stuart England* (New York: G. P. Putnam's Sons, 1959), 272–74.

has every reason to hope and expect from the affection and loyalty of his subjects that none of them, even of those upon whom God may not yet have conferred his divine grace so abundantly as to incline them by that august example to turn to the true faith, will ever fail in the obedience that all people owe to the sovereigns, even of a different religion. Nevertheless, as there are sometimes mischievous and unquiet spirits who seek to disturb the public peace, especially when they can conceal their wicked designs under the plausible excuse of religion, his Majesty of Great Britain, who has nothing more at heart . . . than to confirm the peace which the mildness of his government has gained for his subjects, has concluded that the best means to prevent any alteration in it would be to make himself assured in case of need of the assistance of his most Christian Majesty, who, wishing in this case to give to the lord King of Great Britain an unquestionable proof of the reality of his friendship, and to contribute to the success of so glorious a design, and one of such service not merely to his Majesty of Great Britain but also to the whole Catholic religion . . . promises to give for that purpose to the said lord King of Great Britain the sum of two million *livres tournois*.[67] . . . In addition the said most Christian King binds himself to assist his Majesty of Great Britain in case of need with troops to the number of 6000 foot soldiers, and even to raise and maintain them at his own expense, so far as the said lord King of Great Britain finds need of them for the execution of his design; and the said troops shall be transported by ships of the King of Great Britain to such places and ports as he shall consider most convenient for the good of his service, and from the day of their embarkation shall be paid, as agreed, by his most Christian Majesty, and shall obey the orders of the said lord King of Great Britain. And the time of the said declaration of Catholicism is left entirely to the choice of the said lord King of Great Britain.

3. It has also been agreed . . . that the said most Christian King shall never break or infringe the peace which he has made with Spain. . . .

4. It is also agreed and accepted that if there should hereafter fall to the most Christian King any new titles and rights to the Spanish monarchy, the said lord King of Great Britain shall assist his most Christian Majesty with all his forces . . . to facilitate the acquisition of the said rights. . . .

5. The said lord Kings having each in his own right many more subjects than they would have any need of to justify to the world the resolution they have taken to humble the pride of the States General of the United Provinces of the Low Countries, and to reduce the power of a nation which has so often rendered itself odious by extreme ingratitude

[67] *Livres* of Tours, France, a coin so called in contrast to the *livre* of Paris.

to its own founders and the creators of its republic, and which even has the insolence to aim now at setting itself up as sovereign arbiter and judge of all other potentates, it is agreed, decided and concluded that their Majesties will declare and wage war jointly with all their forces . . . on the said States General of the United Provinces of the Low Countries, and that neither of the said lord Kings will make any treaty of peace, or truce, or suspension of arms with them without the knowledge and consent of the other, as also that all commerce between the subjects of the said lord Kings and those of the said States shall be forbidden, and that the vessels and goods of those who carry on trade in defiance of this prohibition may be seized by the subjects of the other lord King. . . .

6. And for the purpose of waging and conducting the war . . . it is also agreed that his most Christian Majesty will undertake all the expense necessary for setting on foot, maintaining and supporting the operations of the armies required for delivering a powerful attack by land on the strongholds and territory of the said States, the said lord King of Great Britain binding himself only to contribute to the army of the said most Christian King, and to maintain there at his own expense, a body of 6000 infantry. . . .

7. As to what concerns the war at sea, the said lord King of Great Britain shall undertake that burden, and shall fit out at least fifty great ships and ten fire-ships, to which the said most Christian King shall bind himself to add a squadron of thirty good French vessels. . . . And in order that the said lord King of Great Britain may more easily support the expense of the war, his most Christian Majesty binds himself to pay to the said King each year that the said war shall last the sum of three millions of *livres tournois*. . . . And of all the conquests which shall be made from the States General his Majesty of Great Britain shall be content with the following places, viz., the island of Walcheren, Sluys, with the island of Cadsand. . . . And inasmuch as the dissolution of the government of the States General might involve some prejudice to the Prince of Orange, nephew to the King of Great Britain, and also that some fortresses, towns and governments which belong to him are included in the proposed division of the country, it has been determined and concluded that the said lord Kings shall do all they can to secure that the said Prince may find his advantage in the continuation and end of the war, as shall hereafter be provided in separate articles. . . .

196. The Disagreement of King and Commons, 1677

Two years after agreeing to the secret Treaty of Dover, Charles II again involved the realm in another war against the Dutch. It was, however, unpopular, and Parliament, by withholding funds, forced the

King to make peace. There was growing sentiment that France, rather than the Netherlands, was the prime threat to English security and prosperity. In 1677 the House of Commons addressed the King in support of alliances against France, a move resented by Charles as infringing upon the royal prerogative of determining foreign policy.

May it please your most excellent Majesty:

Your Majesty's most loyal and dutiful subjects, the Commons in Parliament assembled, having taken into their serious consideration your Majesty's gracious speech, do beseech your Majesty to believe it is a great affliction to them to find themselves obliged at present to decline the granting your Majesty the supply your Majesty is pleased to demand, conceiving it is not agreeable to the usage of Parliament to grant supplies for maintenance of wars and alliances before they are signified in Parliament; which the two wars against the States of the United Provinces since your Majesty's happy restoration, and the league made with them in January 1668 for preservation of the Spanish Netherlands, sufficiently prove, without troubling your Majesty with instances of greater antiquity. From which usage if we should depart, the precedent might be of dangerous consequence in future times, though your Majesty's goodness gives us great security during your Majesty's reign, which we beseech God long to continue.

This consideration prompted us, in our last address to your Majesty before our late recess, humbly to mention to your Majesty our hopes that before our meeting again your Majesty's alliances might be so fixed as that your Majesty might be graciously pleased to impart them to us in Parliament, that so our earnest desires of supplying your Majesty for prosecuting those great ends we had humbly laid before your Majesty might meet with no impediment or obstruction, being highly sensible of the necessity of supporting as well as making the alliances humbly desired in our former addresses, and which we still conceive so important to the safety of your Majesty and your kingdoms that we cannot, without unfaithfulness to your Majesty and those we represent, omit upon all occasions humbly to beseech your Majesty, as we now do, to enter into a league, offensive and defensive, with the States General of the United Provinces, against the growth and power of the French king, and for the preservation of the Spanish Netherlands, and to make such other alliances with such other of the confederates as your Majesty shall think fit and useful to that end. . . .

SOURCE: *Journals of the House of Commons* (London: Her Majesty's Stationery Office), *IX*, 425–26.

Charles II's Reply – Gentlemen:

Could I have been silent I would rather have chosen to be so than to call to mind things so unfit for you to meddle with as are contained in some part of your address, wherein you have intrenched upon so undoubted a right of the crown that I am confident it will appear in no age (when the sword was not drawn) that the prerogative of making peace and war hath been so dangerously invaded. You do not content yourselves with desiring me to enter into such leagues as may be for the safety of the kingdom, but you tell me what sort of leagues they must be, and with whom; and as your address is worded it is more liable to be understood to be by your leave than your request that I should make such other alliances as I please with other of the confederates. Should I suffer this fundamental power of making peace and war to be so far invaded (though but once) as to have the manner and circumstances of leagues prescribed to me by Parliament, it is plain that no prince or state would any longer believe that the sovereignty of England rests in the crown; nor could I think myself to signify any more to foreign princes than the empty sound of a king. Wherefore you may rest assured that no condition shall make me depart from, or lessen, so essential a part of the monarchy; and I am willing to believe so well of this House of Commons that I am confident these ill consequences are not intended by you.

These are, in short, the reasons why I can by no means approve of your address. And yet, though you have declined to grant me that supply which is so necessary to the ends of it, I do again declare to you that, as I have done all that lay in my power since your last meeting, so I will still apply myself by all means I can to let the world see my care both for the security and satisfaction of my people, although it may not be with those advantages to them which by your assistance I might have procured.

SECTION K

Constitutional Refinements under Charles II

197. The Commons and Money Bills

After the Restoration the House of Commons continued, royalist sentiments notwithstanding, jealously to guard its powers over the purse

---

SOURCE: *Journals of the House of Commons* (London: Her Majesty's Stationery Office), *IX*, 235, 509.

strings. Twice under Charles II it claimed that the Lords had no right to amend money bills: in other words, that the upper house must either accept or reject measures launched in the House of Commons.[68]

Instance of April 13, 1671 – The House then proceeded to the reading the amendments and clauses, sent from the Lords, to the bill for an imposition on foreign commodities, which were once read;

And the first amendments, sent from the Lords, being for changing the proportion of the impositions on white sugars from one penny per pound to halfpenny half farthing, was read the second time and debated.

Resolved, etc., *nemine contradicente,*[69] that in all aids given to the King by the Commons, the rate or tax ought not to be altered by the Lords.

Instance of July 3, 1678 – Mr. Solicitor-General reports from the committee to whom it was, amongst other things, referred, to prepare and draw up a state of the rights of the Commons, in granting of money, a vote agreed by the committee; which he read at his place and afterwards delivered the same in at the Clerk's table, where the same was read, and upon the question agreed, and is as followeth, viz.:

Resolved, etc., that all aids and supplies, and aids to his Majesty in Parliament, are the sole gift of the Commons; and all bills for the granting of any such aids and supplies ought to begin with the Commons; and that it is the undoubted and sole right of the Commons to direct, limit and appoint, in such bills, the ends, purposes, considerations, conditions, limitations and qualifications of such grants, which ought not to be changed or altered by the House of Lords.

198. The Independence of the Jury

In 1670 a judicial decision in Bushell's Case established the independence of juries, in the sense that jurymen were not to be punished for verdicts regarded as wrong by the court. A means of influencing or controlling juries was thus removed, and since then no English juryman has suffered under that charge.

In the present case it is returned that the prisoner [Edward Bushell], being a juryman among others charged at the Sessions Court of the Old

SOURCE: Thomas B. Howell, *Complete Collection of State Trials* (London: R. Bagshaw [etc.], 1809–26), *VI*, 999–1013.

[68] On the Commons and the launching of money bills, see p. 150.
[69] Unanimously.

Bailey to try the issue between the King and Penn[70] and Mead, upon an indictment for assembling unlawfully and tumultuously, did *contra plenam et manifestam evidentiam,*[71] openly given in court, acquit the prisoners indicted, in contempt of the King, etc. . . .

I would know whether anything be more common than for two men, students, barristers or judges, to deduce contrary and opposite conclusions out of the same case in law? And is there any difference that two men should infer distinct conclusions from the same testimony? Is anything more known than that the same author, and place in that author, is forcibly urged to maintain contrary conclusions, and the decision hard, which is in the right? Is anything more frequent in the controversies of religion than to press the same text for opposite tenets? How then comes it to pass that two persons may not apprehend with reason and honesty what a witness, or many, say, to prove in the understanding of one plainly one thing, but in the apprehension of the other clearly the contrary thing? Must therefore one of these merit fine and imprisonment, because he doth that which he cannot otherwise do, preserving his oath and integrity? And this often is the case of the judge and jury.

I conclude therefore that this return, charging the prisoners to have acquitted Penn and Mead against full and manifest evidence, . . . without saying that they did know and believe that evidence to be full and manifest against the indicted persons, is no cause of fine or imprisonment. . . .

We come now to the next part of the return, viz., "That the jury acquitted those indicted against the direction of the court in matter of law, openly given and declared to them in court.". . .

But the reasons are, I conceive, most clear that the judge could not nor can fine and imprison the jury in such cases.

Without a fact agreed, it is as impossible for a judge or any other to know the law relating to that fact or direct concerning it, as to know an accident that hath no subject.

Hence it follows that the judge can never direct what the law is in any matter controverted without first knowing the fact; and then it follows that, without his previous knowledge of the fact, the jury cannot go against his direction in law, for he could not direct.

But the judge, quâ judge, cannot know the fact possibly but from the evidence which the jury have, but . . . he can never know what evidence the jury have, and consequently he cannot know the matter of fact, nor punish the jury for going against their evidence, when he cannot know what their evidence is.

It is true, if the jury were to have no other evidence for the fact but what is deposed in court, the judge might know their evidence and the

70 William Penn, the prominent Quaker and founder of Pennsylvania.

71 Against full and manifest evidence.

fact from it, equally as they, and so direct what the law were in the case, though even then the judge and jury might honestly differ in the result from the evidence, as well as two judges may, which often happens. But the evidence which the jury have of the fact is much other than that[72]. . .

That *Decantatum*[73] in our books, *Ad quaestionem facti non respondent judices; ad quaestionem legis non respondent juratores,*[74] literally taken is true; for if it be demanded, what is the fact, the judge cannot answer it; if it be asked, what is the law in the case, the jury cannot answer it. . . .

199. The Habeas Corpus Act, 1679

Although the writ of habeas corpus, designed to prevent arbitrary imprisonment, had a long history in England before Charles II's time, the government had found ways of circumventing it, particularly in political offenses. The Habeas Corpus Act of 1679 contributed significantly to the liberties of the subject by strengthening the writ's effectiveness.

Whereas great delays have been used by sheriffs, jailers and other officers, to whose custody any of the King's subjects have been committed for criminal or supposed criminal matters, in making returns of writs of habeas corpus to them directed . . . to avoid their yielding obedience to such writs, contrary to their duty and the known laws of the land; whereby many of the King's subjects have been . . . long detained in prison in such cases where by law they are bailable, to their great charge and vexation:

For the prevention whereof and the more speedy relief of all persons imprisoned for any such criminal or supposed criminal matters, be it enacted . . . that, whensoever any person or persons shall bring any habeas corpus directed unto any sheriff or sheriffs, jailer, minister or other person whatsoever for any person in his or their custody, and the said writ shall be served upon the said officer or left at the jail or prison with any of the under-officers . . . , the said officer or officers, his or their under-

SOURCE: *Statutes of the Realm* (London: Her Majesty's Stationery Office, 1810–28), V, 935–37.

[72] It is clear from this passage that at this time all evidence did not have to be introduced in open court.
[73] Something said over and over.
[74] To a question of fact the judges may not respond; to a question of law the jurors may not respond.

officers . . . , shall within three days after the service thereof as aforesaid – unless the commitment aforesaid were for treason or felony plainly and specially expressed in the warrant of commitment – . . . make return of such writ or bring or cause to be brought the body of the party so committed or restrained unto or before the Lord Chancellor or Lord Keeper of the great seal of England for the time being, or the judges or barons[75] of the said court from whence the said writ shall issue, or unto or before such other person . . . before whom the said writ is made returnable according to the command thereof; and shall likewise then certify the true causes of his detainer or imprisonment. . . .

And for the prevention of unjust vexation by reiterated commitments for the same offense, be it enacted . . . that no person or persons, which shall be delivered or set at large upon any habeas corpus, shall at any time hereafter be again imprisoned or committed for the same offense by any person or persons whatsoever other than by the legal order and process of such court wherein he or they shall be bound by recognizance to appear, or other court having jurisdiction of the cause; . . . and if any other person or persons shall knowingly contrary to this Act recommit or imprison . . . for the same offense or pretended offense any person or persons delivered or set at large as aforesaid, . . . then he or they shall forfeit to the prisoner or party grieved the sum of five hundred pounds . . .

Provided always . . . that, if any person or persons shall be committed for high treason or felony, . . . [and] shall not be indicted sometime in the next term sessions of oyer and terminer or general jail-delivery after such commitment, it shall . . . be lawful . . . for the judges of the Court of King's Bench and justices of oyer and terminer or general jail-delivery . . . to set at liberty the prisoner upon bail, unless it appear to the judges and justices upon oath made that the witnesses for the King could not be produced the same term sessions or general jail-delivery. And if any person or persons, committed as aforesaid, . . . shall not be indicted and tried the second term sessions . . . after his commitment, or upon his trial shall be acquitted, he shall be discharged from his imprisonment. . . .

[75] The judges of the Court of Exchequer were known as barons.

SECTION L

Political Theory on the Eve of the Revolution

200. Robert Filmer's *Patriarcha*

On the eve of the Revolution English political ideas ranged from uncompromising theories of divine right to various concepts of popular sovereignty. The most noteworthy work of this period in support of the divine right theory of kingship was Sir Robert Filmer's *Patriarcha, or the Natural Power of Kings,* first published in 1680, though written a generation earlier. Filmer's main contention was that monarchical power is patriarchal or paternal in essence, and hence natural and bestowed of God. The first of John Locke's *Treatises of Government* was written as a rebuttal.

Within the last hundred years many of the schoolmen and other divines have published and maintained an opinion that: "Mankind is naturally endowed and born with freedom from all subjection, and at liberty to choose what form of government it please, and that the power which any one man hath over others was at the first bestowed according to the discretion of the multitude."

This tenet was first hatched in the schools for good divinity, and hath been fostered by succeeding papists. The divines of the reformed churches have entertained it, and the common people everywhere tenderly embrace it . . . , never remembering that the desire of liberty was the first cause of the fall of Adam.

But howsoever this opinion hath of late obtained great reputation, yet it is not to be found in the ancient fathers and doctors of the primitive church. It contradicts the doctrine and history of the Holy Scriptures, the constant practice of all ancient monarchies and the very principles of the law of nature. It is hard to say whether it be more erroneous in divinity or dangerous in policy.

Upon the grounds of this doctrine, both Jesuits and some zealous favorers of the Geneva discipline have built a perilous conclusion, which is "that the people or multitude have power to punish or deprive the prince if he transgress the laws of the kingdom." Witness Parsons[76] and

SOURCE: Peter Laslett (ed.), *Patriarcha and Other Political Works of Sir Robert Filmer* (Oxford: Basil Blackwell, 1949), 53–54, 62–63.

[76] Robert Parsons (1546–1610), an English Jesuit.

Buchanan.[77]. . . Cardinal Bellarmino[78] and Mr. Calvin both look asquint this way. . . .

In all kingdoms or commonwealths in the world, whether the prince be the supreme father of the people or but the true heir of such a father, or whether he come to the crown by usurpation, or by election of the nobles or of the people, or by any other way whatsoever, or whether some few or a multitude govern the commonwealth, yet still the authority that is in any one, or in many, or in all of these, is the only right and natural authority of a supreme father. There is and always shall be continued to the end of the world a natural right of a supreme father over every multitude, although, by the secret will of God, many at first do most unjustly obtain the exercise of it.

To confirm this natural right of regal power, we find in the Decalogue that the law which enjoins obedience to kings is delivered in the terms of "Honor thy father," as if all power were originally in the father. If obedience to parents be immediately due by a natural law, and subjection to princes but by the mediation of a human ordinance, what reason is there that the law of nature should give place to the laws of men, as we see the power of the father over his child gives place and is subordinate to the power of the magistrate? . . .

If we compare the natural duties of a father with those of a king, we find them to be all one, without any difference at all but only in the latitude or extent of them: as the father over one family, so the king, as father over many families, extends his care to preserve, feed, clothe, instruct and defend the whole commonwealth. His wars, his peace, his courts of justice, and all his acts of sovereignty, tend only to preserve and distribute to every subordinate and inferior father, and to their children, their rights and privileges, so that all the duties of a king are summed up in an universal fatherly care of his people. . . .

201. John Locke's *Second Treatise of Government*

Locke's treatises on government, though written several years before the Revolution of 1688, came to be regarded as a defense of that crucial event. The *Second Treatise* in particular exercised an extraordinary influence on English and American political thought for generations. In the following passages from this work some of Locke's

---

SOURCE: John Locke, *Two Treatises of Government* (London: Awnsham and John Churchill, 1698), 231–32, 267–68, 286, 337–38.

---

[77] George Buchanan (1506–1582), the Scottish historian and scholar.
[78] Cardinal Roberto Bellarmino (1542–1621), a Jesuit theologian.

principal ideas are set forth, as the social contract, the supreme power of the legislature (based on the sovereignty of the people), and the responsibility of legislature and executive to provide good government.

Wherever . . . any number of men are so united into one society as to quit everyone his executive power of the law of nature, and to resign it to the public, there and there only is a political or civil society. And this is done wherever any number of men, in the state of nature, enter into society to make one people one body politic under one supreme government; or else, when anyone joins himself to and incorporates with any government already made. For hereby he authorizes the society, or which is all one, the legislative thereof, to make laws for him as the public good of the society shall require, to the execution whereof his own assistance (as to his own decrees) is due. And this puts men out of a state of nature into that of a commonwealth, by setting up a judge on earth with authority to determine all the controversies and redress the injuries that may happen to any member of the commonwealth, which judge is the legislative or magistrates appointed by it. And wherever there are any number of men, however associated, that have no such decisive power to appeal to, there they are still in the state of nature. . . .

The great end of men's entering into society being the enjoyment of their properties in peace and safety, and the great instrument and means of that being the laws established in that society, the first and fundamental positive law of all commonwealths is the establishing of the legislative power, as the first and fundamental natural law which is to govern even the legislative itself, is the preservation of the society and (as far as will consist with the public good) of every person in it. This legislative is not only the supreme power of the commonwealth, but sacred and unalterable in the hands where the community have once placed it. Nor can any edict of anybody else, in what form soever conceived, or by what power soever backed, have the force and obligation of a law which has not its sanction from that legislative which the public has chosen and appointed; for without this the law could not have that which is absolutely necessary to its being a law, the consent of the society, over whom nobody can have a power to make laws but by their own consent and by authority received from them; and therefore all the obedience, which by the most solemn ties anyone can be obliged to pay, ultimately terminates in this supreme power, and is directed by those laws which it enacts. Nor can any oaths to any foreign power whatsoever, or any domestic subordinate power, discharge any member of the society from his obedience to the legislative, acting pursuant to their trust, nor oblige him to any obedience contrary to the laws so enacted or farther than they do allow, it being ridiculous to imagine one can be tied ultimately to obey any power in the society which is not the supreme. . . .

It may be demanded here, what if the executive power, being possessed of the force of the commonwealth, shall make use of that force to hinder the meeting and acting of the legislative, when the original constitution or public exigencies require it? I say, using force upon the people, without authority, and contrary to the trust put in him that does so, is a state of war with the people, who have a right to reinstate their legislative in the exercise of their power. For having erected a legislative with an intent they should exercise the power of making laws, either at certain set times or when there is need of it, when they are hindered by any force from what is so necessary to the society, and wherein the safety and preservation of the people consists, the people have a right to remove it by force. In all states and conditions the true remedy of force without authority is to oppose force to it. The use of force without authority always puts him that uses it into a state of war as the aggressor, and renders him liable to be treated accordingly. . . .

The reason why men enter into society is the preservation of their property; and the end why they choose and authorize a legislative is that there may be laws made and rules set as guards and fences to the properties of all the members of the society, to limit the power and moderate the dominion of every part and member of the society. For since it can never be supposed to be the will of the society that the legislative should have a power to destroy that which everyone designs to secure by entering into society, and for which the people submitted themselves to the legislators of their own making: whenever the legislators endeavor to take away and destroy the property of the people, or to reduce them to slavery under arbitrary power, they put themselves into a state of war with the people, who are thereupon absolved from any farther obedience, and are left to the common refuge which God hath provided for all men against force and violence. Whensoever, therefore, the legislative shall transgress this fundamental rule of society, and either by ambition, fear, folly or corruption endeavor to grasp themselves, or put into the hands of any other, an absolute power over the lives, liberties and estates of the people, by this breach of trust they forfeit the power the people had put into their hands for quite contrary ends, and it devolves to the people, who have a right to assume their original liberty, and by the establishment of a new legislative (such as they shall think fit), provide for their own safety and security, which is the end for which they are in society. What I have said here concerning the legislative in general holds true also concerning the supreme executor, who having a double trust put in him, both to have a part in the legislative and the supreme execution of the law, acts against both when he goes about to set up his own arbitrary will as the law of the society. . . .

SECTION M

Controversial Issues under James II

202. The Appointment of Catholic Army Officers, 1685

Upon ascending the throne, James II sought at once to appoint Roman Catholics to civil and military offices, regardless of the Test Act. This raised the constitutional question of the King's right to dispense with the law. James's awareness of this crucial issue is shown in the following speech to Parliament, in which he also touches on the need for a regular army, a most unpopular theme.

My lords and gentlemen, after the storm that seemed to be coming upon us when we parted last, I am glad to meet you all again in so great peace and quietness. God almighty be praised, by whose blessing that rebellion[79] was suppressed. But when we reflect what an inconsiderable number of men began it, and how long they carried [it] on without any opposition, I hope everybody will be convinced that the militia, which has hitherto been so much depended on, is not sufficient for such occasions, and that there is nothing but a good force of well-disciplined troops in constant pay that can defend us from such as, either at home or abroad, are disposed to disturb us. And in truth, my concern for the peace and quiet of my subjects, as well as for the safety of the government, made me think it necessary to increase the number to the proportion I have done. This I owed as well to the honor as the security of the nation, whose reputation was so infinitely exposed to all our neighbors, by having so evidently lain open to this late wretched attempt, that it is not to be repaired without keeping such a body of men on foot that none may ever have the thought again of finding us so miserably unprovided.

SOURCE: *Journals of the House of Commons* (London: Her Majesty's Stationery Office), *IX*, 756.

---

[79] Monmouth's Rebellion, a rising on behalf of the Protestant Duke of Monmouth, an illegitimate son of Charles II.

It is for the support of this great charge, which is now more than double to what it was, that I ask your assistance in giving me a supply answerable to the expense it brings along with it. And I cannot doubt but what I have begun, so much for the honor and defense of the government, will be continued by you with all the cheerfulness that is requisite for a work of so great importance.

Let no man take exception that there are some officers in the army not qualified, according to the late Test Acts, for their employment. The gentlemen, I must tell you, are most of them well known to me, and having formerly served with me in several occasions, and always approved the loyalty of their principles by their practice, I think fit now to be employed under me. And I deal plainly with you, that after having had the benefit of their service in such time of need and danger, I will neither expose them to disgrace, nor myself to want of them, if there should be another rebellion to make them necessary for me.

I am afraid some men may be so wicked to hope and expect that a difference may happen between you and me upon this occasion. But when you consider what advantages have arisen to us in a few months by the good understandings we have hitherto had; what wonderful effects it has already produced in the change of the whole scene of affairs abroad, so much more to the honor of this nation and the figure it ought to make in the world; and that nothing can hinder a further progress in this way, to all our satisfactions, but fears and jealousies amongst ourselves, I will not apprehend that such a misfortune can befall us as a division, or but a coldness, between me and you, nor that anything can shake you in your steadiness and loyalty to me who, by God's blessing, will ever make you returns of all kindness and protection, with a resolution to venture even my own life in the defense of the true interest of this kingdom.

203. The Decision in Godden vs. Hales, 1686

Rebuffed by Parliament, King James resorted to a judicial decision to test the legality of the dispensing power. In a suit brought by collusive action against Sir Edward Hales, a Catholic convert who had accepted the military governorship of Dover Castle without taking the required oaths, the judges, eleven to one, supported the royal position, on grounds set forth by Chief Justice Herbert.

In the case of Godden and Hales, wherein the defendant pleads a dispensation from the King, it is doubted whether or no the King had such

---

SOURCE: Thomas B. Howell, *Complete Collection of State Trials* (London: R. Bagshaw [etc.], 1809–26), *XI*, 1197–99.

a prerogative. Truly, upon the argument before us, it appeared as clear a case as ever came before this court; but, because men fancy I know not what difficulty where really there is none, we were willing to give so much countenance to the question in the case as to take the advice of all the judges in England.[80] They were all assembled at Serjeants' Inn . . .

And I must tell you that there were ten upon the place that clearly delivered their opinions. . . . My brother[81] Powell said he was inclined to be of the same opinion, but he would rather have some more time to consider of it. But he has since sent by my brother Holloway to let us know that he does concur with us. To these eleven judges there is one dissenter, brother Street, who yet continues his opinion that the King cannot dispense in this case. But that's the opinion of one single judge against the opinion of eleven.

We were satisfied in our judgments before and, having the concurrence of eleven out of twelve, we think we may very well declare the opinion of the court to be that the King may dispense in this case. And the judges go upon these grounds: that the kings of England are sovereign princes; that the laws of England are the king's laws; that therefore 'tis an inseparable prerogative in the kings of England to dispense with penal laws in particular cases and upon particular necessary reasons; that of those reasons and those necessities, the king himself is sole judge, and then, which is consequent upon all, that this is not a trust invested in, or granted to, the king by the people, but the ancient remains of the sovereign power and prerogative of the kings of England; which never yet was taken from them, nor can be. And therefore, such a dispensation appearing upon record to come [in] time enough to save him from the forfeiture, judgment ought to be given for the defendant.

204. The Petition of the Bishops, 1688

Like Charles II, King James sought to suspend the penal clauses in the laws on religion, and he too ran into strong opposition. When in 1688 he ordered clergymen to proclaim his second Declaration of Indulgence, seven bishops refused to comply. In a petition to the King they protested against the Declaration as illegal, along lines already taken by the Commons in 1672.

SOURCE: Thomas B. Howell, *Complete Collection of State Trials* (London: R. Bagshaw [etc.], 1809–26, *XII*, 318–19.

---

80 I.e., all twelve judges of the principal common law courts.
81 Colleague.

To the King's most excellent Majesty: the humble petition of William, Archbishop of Canterbury, and of divers of the suffragan bishops of that province, now present with him, in behalf of themselves and others of their absent brethren and of the clergy of their respective dioceses, humbly showeth:

That the great averseness they find in themselves to the distributing and publishing in all their churches your Majesty's late declaration for liberty of conscience proceedeth neither from any want of duty and obedience to your Majesty, our holy mother the Church of England being both in her principles and in her constant practice unquestionably loyal, nor yet from any want of due tenderness to dissenters, in relation to whom they are willing to come to such a temper as shall be thought fit when the matter shall be considered and settled in Parliament and Convocation, but, amongst many other considerations, from this especially, because that declaration is founded upon such a dispensing power as hath been often declared illegal in Parliament, and particularly in the years 1662 and 1672, and in the beginning of your Majesty's reign, and is a matter of so great moment and consequence to the whole nation, both in church and state, that your petitioners cannot in prudence, honor or conscience so far make themselves parties to it as the distribution of it all over the nation and the solemn publication of it once and again, even in God's house and in the time of his divine service, must amount to, in common and reasonable construction.

Your petitioners therefore most humbly and earnestly beseech your Majesty that you will be graciously pleased not to insist upon their distributing and reading your Majesty's said declaration. . . .

SECTION N

The Glorious Revolution

205. The Invitation to William of Orange, 1688

Growing opposition to James II's political methods and Catholic proclivities was climaxed in June 1688 by the birth of an heir to the

---

SOURCE: John Dalrymple, *Memoirs of Great Britain and Ireland* (London: W. Strahan and T. Cadell [etc.], 1771–88), *II*, 228–31.

throne. Fearful that James's system would be continued by a Catholic line of kings, seven prominent Englishmen, of both political parties, issued an invitation to William of Orange (the husband of James's daughter Mary) to undertake an invasion in support of English liberties.

We have great satisfaction to find . . . that your Highness is sc ready and willing to give us such assistance as they [secret agents] have related to us. We have great reason to believe we shall be every day in a worse condition than we are, and less able to defend ourselves, and therefore we do earnestly wish we might be so happy as to find a remedy before it be too late for us to contribute to our own deliverance. But although these be our wishes, yet we will by no means put your Highness into any expectations which may misguide your own councils in this matter; so that the best advice we can give is to inform your Highness truly both of the state of things here at this time and of the difficulties which appear to us.

As to the first, the people are so generally dissatisfied with the present conduct of the government in relation to their religion, liberties and properties (all which have been greatly invaded), and they are in such expectation of their prospects being daily worse, that your Highness may be assured there are nineteen parts of twenty of the people throughout the kingdom who are desirous of a change, and who, we believe, would willingly contribute to it, if they had such a protection to countenance their rising as would secure them from being destroyed before they could get to be in a posture able to defend themselves. It is no less certain that much the greatest part of the nobility and gentry are as much dissatisfied, although it be not safe to speak to many of them beforehand; and there is no doubt but some of the most considerable of them would venture themselves with your Highness at your first landing, whose interests would be able to draw great numbers to them whenever they could protect them and the raising and drawing men together. And if such a strength could be landed as were able to defend itself and them till they could be got together into some order, we make no question but that strength would quickly be increased to a number double to the army here, although their army should all remain firm to them; whereas we do upon very good grounds believe that their army then would be very much divided among themselves, many of the officers being so discontented that they continue in their service only for a subsistence (besides that some of their minds are known already), and very many of the common soldiers do daily show such an aversion to the popish religion that there is the greatest probability imaginable of great numbers of deserters which would come from them should there be such an occasion; and amongst the seamen it is almost certain there is not one in ten who would do them any service in such a war.

Besides all this, we do much doubt whether this present state of things will not yet be much changed to the worse before another year, by a great alteration which will probably be made both in the officers and soldiers of the army, and by such other changes as are not only to be expected from a packed Parliament, but what the meeting of any Parliament (in our present circumstances) may produce against those who will be looked upon as principal obstructors of their proceedings there, it being taken for granted that if things cannot then be carried to their wishes in a parliamentary way other measures will be put in execution by more violent means; and although such proceedings will then heighten the discontents, yet such courses will probably be taken at that time as will prevent all possible means of relieving ourselves.

These considerations make us of opinion that this is a season in which we may more probably contribute to our own safeties than hereafter (although we must own to your Highness there are some judgments differing from ours in this particular), insomuch that if the circumstances stand so with your Highness that you believe you can get here time enough, in a condition to give assistance this year sufficient for a relief under these circumstances which have been now represented, we who subscribe this will not fail to attend your Highness upon your landing and to do all that lies in our power to prepare others to be in as much readiness as such an action is capable of, where there is so much danger in communicating an affair of such a nature till it be near the time of its being made public. But, as we have already told your Highness, we must also lay our difficulties before your Highness, which are chiefly that we know not what alarm your preparations for this expedition may give, or what notice it will be necessary for you to give the States beforehand, by either of which means their intelligence or suspicions here may be such as may cause us to be secured before your landing. And we must presume to inform your Highness that your compliment upon the birth of the child[82] (which not one in a thousand here believes to be the Queen's) has done you some injury, the false imposing of that upon the Princess[83] and the nation being not only an infinite exasperation of people's minds here, but being certainly one of the chief causes upon which the declaration of your entering the kingdom in a hostile manner must be founded on your part, although many other reasons are to be given on ours.

If upon a due consideration of all these circumstances your Highness shall think fit to adventure upon the attempt, or at least to make such

[82] The birth of Prince James, in June 1688, to James II's consort, Mary of Modena.
[83] Princess Mary, elder daughter of James II, who prior to the birth of Prince James had been heir apparent.

preparations for it as are necessary (which we wish you may), there must be no more time lost in letting us know your resolution concerning it, and in what time we may depend that all the preparations will be ready, as also whether your Highness does believe the preparations can be so managed as not to give them warning here, both to make them increase their force and to secure those they shall suspect would join with you. We need not say anything about ammunition, artillery, mortar pieces, spare arms, etc., because if you think fit to put anything in execution you will provide enough of these kinds, and will take care to bring some good engineers with you; and we have desired Mr. H. to consult you about all such matters, to whom we have communicated our thoughts in many particulars too tedious to have been written, and about which no certain resolutions can be taken till we have heard again from your Highness. . . .

206. The Invasion of William of Orange, 1688

Having assembled his forces, William of Orange set sail for England and, evading the English fleet, landed in the southwestern part of the country. Gilbert Burnet, later Bishop of Salisbury, who accompanied William, has left us this account of the invasion.

The Prince desired me to go along with him as his chaplain, to which I very readily agreed; for, being fully satisfied in my conscience that the undertaking was lawful and just, and having had a considerable hand in advising the whole progress of it, I thought it would have been an unbecoming fear in me to have taken care of my own person when the Prince was venturing his, and the whole was now to be put to hazard. . . .

As Sidney brought over letters . . . inviting the Prince to come over to save and rescue the nation from ruin, and assuring him that they wrote that which was the universal sense of all the wise and good men in the nation, so they also sent over with him a scheme of advices. They advised his having a great fleet but a small army; they thought it should not exceed six or seven thousand men. They apprehended that an ill use might be made of it, if he brought over too great an army of foreigners, to infuse in people a jealousy that he designed a conquest. They advised his landing in the north, either in Burlington Bay or a little below Hull; Yorkshire abounded in horse, and the gentry were generally well affected, even to zeal, for the design; the country was plentiful and the roads were good till within fifty miles of London. . . .

SOURCE: Gilbert Burnet, *History of His Own Time* (London: T. Ward [etc.], 1724–34), *I*, 776–89.

When these things were laid before the Prince, he said he could by no means resolve to come over with so small a force; he could not believe what they suggested concerning the King's army's being disposed to come over to him, nor did he reckon so much as they did on the people of the country's coming in to him. He said he could trust to neither of these; he could not undertake so great a design, the miscarriage of which would be the ruin both of England and Holland, without such a force as he had reason to believe would be superior to the King's own, though his whole army should stick to him. . . .

When these advices were proposed to Herbert and the other seamen, they opposed the landing in the north vehemently. They said no seamen had been consulted in that. The north coast was not fit for a fleet to ride in, in an east wind, which it was to be expected in winter might blow so fresh that it would not be possible to preserve the fleet; and if the fleet was left there the Channel was open for such forces as might be sent from France. The Channel was the safer sea for the fleet to ride in, as well as to cut off the assistance from France. . . .

In the beginning of October the troops marched from Nimeguen and were put on board in the Zuyder Sea, where they lay above ten days before they could get out of the Texel. Never was so great a design executed in so short a time. A transport fleet of five hundred vessels was hired in three days' time. All things, as soon as they were ordered, were got to be so quickly ready that we were amazed at the dispatch. It is true, some things were wanting and some things had been forgot. But when the greatness of the equipage was considered, together with the secrecy with which it was to be conducted till the whole design was to be avowed, it seemed much more strange that so little was wanting or that so few things had been forgot. . . .

On the first of November O.S.[84] we sailed out with the evening tide, but made little way that night, that so our fleet might come out and move in order. We tried next day till noon, if it was possible to sail northward, but the wind was so strong and full in the east that we could not move that way. About noon the signal was given to steer westward. This wind not only diverted us from that unhappy course, but it kept the English fleet in the river, so that it was not possible for them to come out, though they were come down as far as to the Gunfleet. By this means we had the sea open to us, with a fair wind and a safe navigation. On the third we passed between Dover and Calais, and before night came in sight of the Isle of Wight. . . . Torbay was thought the best place for our great fleet to lie in, and it was resolved to land the army where it could be best done near it; reckoning that being at such a distance from London we

[84] The Old Style, or Julian, calendar still in use in England. It ran several days behind the New Style, or Gregorian, calendar used in most continental countries.

could provide ourselves with horses and put everything in order before the King could march his army towards us, and that we should lie some time at Exeter for the refreshing our men . . .

And then . . . a soft and happy gale of wind carried in the whole fleet in four hours' time into Torbay. Immediately as many landed as conveniently could. As soon as the Prince and Marshal Schomberg got to shore they were furnished with such horses as the village of Broxholme could afford, and rode up to view the grounds, which they found as convenient as could be imagined for the foot in that season. It was not a cold night; otherwise the soldiers, who had been kept warm aboard, might have suffered much by it. As soon as I landed I made what haste I could to the place where the Prince was, who took me heartily by the hand and asked me if I would not now believe predestination. I told him I would never forget that providence of God, which had appeared so signally on this occasion. He was cheerfuller than ordinary. Yet he returned soon to his usual gravity. . . .

207. The Bill of Rights, 1689

Unable to employ his army effectively against William of Orange, James fled the realm, taking refuge in France. A convention Parliament, similar to that of 1660, was called to work out the political future of the country. Many reforms were suggested. But, in the interest of a speedy reestablishment of regular governmental authority, Parliament drew up a relatively simple settlement, enacted into law as the Bill of Rights, tendering the crown to William and Mary and embodying certain popular rights and limitations upon the royal power as conditions under which the new sovereigns might occupy the throne. Thus Parliament again – and this time conclusively – demonstrated its superiority over divine right kingship and royal irresponsibility.

Whereas the . . . late King James the Second having abdicated the government and the throne being thereby vacant, his Highness the Prince of Orange (whom it has pleased almighty God to make the glorious instrument of delivering this kingdom from popery and arbitrary power) did (by the advice of the Lords Spiritual and Temporal and divers principal persons of the Commons) cause letters to be written to the Lords Spiritual and Temporal being Protestants, and other letters to the several counties, cities, universities, boroughs and cinque ports, for the choosing of such persons to represent them as were of right to be sent to Parliament, to meet and sit at Westminster upon the two and twentieth

---

SOURCE: *Statutes of the Realm* (London: Her Majesty's Stationery Office, 1810–28), *VI*, 143–45.

day of January in this year, in order to such an establishment as that their religion, laws and liberties might not again be in danger of being subverted, upon which letters elections having been accordingly made.

And thereupon the said Lords Spiritual and Temporal and Commons, . . . being now assembled in a full and free representative of this nation, taking into their most serious consideration the best means for attaining the ends aforesaid, do in the first place (as their ancestors in like case have usually done) for the vindicating and asserting their ancient rights and liberties declare:

That the pretended power of suspending of laws or the execution of laws by regal authority without consent of Parliament is illegal.

That the pretended power of dispensing with laws or the execution of laws by regal authority, as it has been assumed and exercised of late, is illegal.

That the commission for erecting the late Court of Commissioners for Ecclesiastical Causes,[85] and all other commissions and courts of like nature, are illegal and pernicious.

That levying money for or to the use of the crown by pretense of prerogative without grant of Parliament, for longer time or in other manner than the same is or shall be granted, is illegal.

That it is the right of the subjects to petition the King, and all commitments and prosecutions for such petitioning are illegal.

That the raising or keeping a standing army within the kingdom in time of peace, unless it be with consent of Parliament, is against law.

That the subjects which are Protestants may have arms for their defense, suitable to their conditions and as allowed by law.

That election of members of Parliament ought to be free.

That the freedom of speech and debates or proceedings in Parliament ought not to be impeached or questioned in any court or place out of Parliament.

That excessive bail ought not to be required, nor excessive fines imposed, nor cruel and unusual punishments inflicted.

That jurors ought to be duly impanelled and returned, and jurors which pass upon men in trials for high treason ought to be freeholders.

That all grants and promises of fines and forfeitures of particular persons before conviction are illegal and void.

And that for redress of all grievances, and for the amending, strengthening and preserving of the laws, Parliament ought to be held frequently.

And they do claim, demand and insist upon all and singular the premises as their undoubted rights and liberties, and that no declarations, judgments, doings or proceedings to the prejudice of the people in any of

[85] A court set up in 1686, bearing a resemblance to the Court of High Commission, abolished in 1641.

the said premises ought in any wise to be drawn hereafter into consequence or example; to which demand of their rights they are particularly encouraged by the declaration of his Highness the Prince of Orange as being the only means for obtaining a full redress and remedy therein. Having therefore an entire confidence that his said Highness . . . will perfect the deliverance so far advanced by him, and will still preserve them from the violation of their rights which they have here asserted, and from all other attempts upon their religion, rights and liberties, the said Lords Spiritual and Temporal and Commons assembled at Westminster do resolve that William and Mary, Prince and Princess of Orange, be and be declared King and Queen of England, France and Ireland and the dominions thereunto belonging, to hold the crown and royal dignity of the said kingdoms and dominions to them . . . during their lives and the life of the survivor of them, and that the sole and full exercise of the regal power be only in and executed by the said Prince of Orange in the names of the said Prince and Princess during their joint lives, and after their deceases the said crown and royal dignity . . . to be to the heirs of the body of the said Princess, and for default of such issue to the Princess Anne of Denmark[86] and the heirs of her body, and for default of such issue to the heirs of the body of the said Prince of Orange. And the Lords Spiritual and Temporal and Commons do pray the said Prince and Princess to accept the same accordingly.

And that the oaths hereafter mentioned be taken by all persons of whom the oaths of allegiance and supremacy might be required by law, instead of them; and that the said oaths of allegiance and supremacy be abrogated.

> I, A. B., do sincerely promise and swear that I will be faithful and bear true allegiance to their Majesties King William and Queen Mary. So help me God.
>
> I, A. B., do swear that I do from my heart abhor, detest and abjure as impious and heretical this damnable doctrine and position that princes excommunicated or deprived by the Pope or any authority of the See of Rome may be deposed or murdered by their subjects or any other whatsoever. And I do declare that no foreign prince, person, prelate, state or potentate hath or ought to have any jurisdiction, power, superiority, pre-eminence or authority, ecclesiastical or spiritual, within this realm. So help me God.

Upon which their said Majesties did accept the crown and royal dignity of the Kingdom of England, France and Ireland, and the dominions

[86] James II's younger daughter, married to Prince George of Denmark.

thereunto belonging, according to the resolution and desire of the said Lords and Commons contained in the said declaration. And thereupon their Majesties were pleased that the said Lords . . . and Commons, being the two houses of Parliament, should continue to sit, and with their Majesties' royal concurrence make effectual provision for the settlement of the religion, laws and liberties of this kingdom, so that the same for the future might not be in danger again of being subverted, to which the said Lords . . . and Commons did agree, and proceed to act accordingly. Now in pursuance of the premises the said Lords . . . and Commons . . . do pray that it may be declared and enacted that all and singular the rights and liberties asserted and claimed in the said declaration are the true, ancient and indubitable rights and liberties of the people of this kingdom, and so shall be esteemed, allowed, adjudged, deemed and taken to be; and that all and every the particulars aforesaid shall be firmly and strictly holden and observed as they are expressed in the said declaration, and all officers and ministers whatsoever shall serve their Majesties and their successors according to the same in all times to come. And the said Lords . . . and Commons . . . do hereby recognize, acknowledge and declare that King James II having abdicated the government, and their Majesties having accepted the crown and royal dignity as aforesaid, their said Majesties did become . . . and of right ought to be by the laws of this realm our sovereign liege lord and lady, King and Queen of England, France and Ireland and the dominions thereunto belonging, in and to whose princely persons the royal state, crown and dignity of the said realms with all honors, styles, titles, regalities, prerogatives, powers, jurisdictions and authorities to the same belonging and appertaining are most fully, rightfully and entirely invested and incorporated, united and annexed. . . . And the said Lords . . . and Commons do in the name of all the people . . . most humbly and faithfully submit themselves, their heirs and posterities forever, and do faithfully promise that they will stand to, maintain and defend their said Majesties, and also the limitation and succession of the crown herein specified and contained, to the utmost of their powers and their lives and estates against all persons whatsoever that shall attempt anything to the contrary. And whereas it has been found by experience that it is inconsistent with the safety and welfare of this Protestant kingdom to be governed by a popish prince, or by any King or Queen marrying a papist, the said Lords . . . and Commons . . . do further pray that it may be enacted that all and every person and persons that is, are or shall be reconciled to or shall hold communion with the See or Church of Rome, or shall profess the popish religion, or shall marry a papist, shall be excluded and be forever incapable to inherit, possess or enjoy the crown and government of this realm . . . or to have, use or exercise any regal power, authority or jurisdiction within the same; and in all and every such case or cases the people of these realms shall be and are hereby absolved of their allegiance; and the said crown and government shall from time to time descend to and be enjoyed by such person or persons being Protestants as should have inherited and enjoyed the same in case the said person or persons so reconciled, holding communion or professing or marrying as aforesaid were naturally dead; and that every King and Queen of this realm who at any time hereafter shall come to and succeed in the imperial crown of this kingdom shall on the first day of the meeting of the first Parliament next after his or her coming to the crown, sitting in his or her throne in the House of Peers in the presence of the Lords and Commons therein assembled, or at his or her coronation before such person or persons who shall administer the coronation oath to him or her . . . make, subscribe and audibly repeat the declaration[87] mentioned in the statute made in the thirtieth year of the reign of King Charles II entitled, An Act for the More Effectual Preserving the King's Person and Government by Disabling Papists from Sitting in Either House of Parliament. . . .

SECTION O

The Economy of Later Stuart England

208. The Navigation Act, 1660

At the end of the seventeenth century most of the Thirteen Colonies had come into being. Although the mother country's direct political control was for the most part rather loose, by 1651 a Navigation Act made it clear that she intended to exercise a firm supervision, along mercantilist lines, over the trade that was gradually being developed. This measure was reenacted with the restoration of the monarchy. Its underlying principles were generally upheld in England until the nineteenth century, despite attacks (particularly Adam Smith's, in his *Wealth of Nations*) on the mercantile system.

SOURCE: *Statutes of the Realm* (London: Her Majesty's Stationery Office, 1810–28), V, 246–47.

---

87 A declaration against the doctrine of transubstantiation, and against the invocation or adoration of the Virgin Mary and other saints.

For the increase of shipping and encouragement of navigation of this nation, wherein, under the good providence and the protection of God, the wealth, safety and strength of this kingdom is so much concerned, be it enacted . . . that from and after the first day of December [1660] . . . no goods or commodities whatsoever shall be imported into or exported out of any lands, islands, plantations or territories to his Majesty belonging . . . in Asia, Africa or America, in any other ship or ships, vessel or vessels whatsoever, but in such ships or vessels as do truly and without fraud belong only to the people of England or Ireland, dominion of Wales or town of Berwick-upon-Tweed, or are of the build of and belonging to any of the said lands, islands, plantations or territories as the proprietors and right owners thereof, and whereof the master and three fourths of the mariners at least are English, under the penalty of the forfeiture and loss of all the goods and commodities . . . , as also of the ship or vessel with all its guns, furniture, tackle, ammunition and apparel. . . .

And it is further enacted . . . that no goods or commodities whatsoever of the growth, production or manufacture of Africa, Asia or America, or of any part thereof . . . be imported into England, Ireland or Wales, island of Guernsey or Jersey or town of Berwick-upon-Tweed, in any other ship or ships than [as above specified] . . . under the penalty of forfeiture. . . .

And it is further enacted . . . that no goods or commodities that are of foreign growth, production or manufacture, and which are to be brought into England, Ireland, Wales, the islands of Guernsey and Jersey or town of Berwick-upon-Tweed in English-built shipping, or other shipping belonging to some of the aforesaid places, and navigated by English mariners as above-said, shall be shipped or brought from any other place or places, country or countries, but only from those of their said growth, production or manufacture, or from those ports where the said goods and commodities can only [be] or are or usually have been first shipped for transportation. . . .

209. Changes in the Poor Law, 1662

Since the days of the Tudors, individual parishes had been made responsible for the relief of the poor.[88] In 1662 the authorities of London, concerned by the influx of the poor to the metropolis, promoted an act which required local officials to prevent the movement

---

SOURCE: *Statutes of the Realm* (London: Her Majesty's Stationery Office, 1810–28), V, 401.

---

[88] See p. 225.

of those likely to require relief, from their native parishes. Thus the mobility of labor, and the freedom of the poor, were limited.

Whereas the necessity, number and continual increase of the poor . . . through the whole Kingdom of England and dominion of Wales is very great and exceeding burdensome, being occasioned by reason of some defects in the law concerning the settling of the poor, and for want of a due provision of the regulations of relief and employment in such parishes or places where they are legally settled, which does force many to turn incorrigible rogues and others to perish for want, together with the neglect of the faithful execution of such laws and statutes as have formerly been made for the apprehending of rogues and vagabonds and for the good of the poor; for remedy whereof, and for the preventing the perishing of any of the poor, whether young or old, for want of such supplies as are necessary, may it please your most excellent Majesty that it may be enacted . . . that whereas by reason of some defects in the poor law people are not restrained from going from one parish to another, and therefore do endeavor to settle themselves in those parishes where there is the best stock,[89] the largest commons or wastes to build cottages, and the most woods for them to burn and destroy, and when they have consumed it, then to another parish and at last become rogues and vagabonds, to the great discouragement of parishes to provide stocks where it is liable to be devoured by strangers; be it therefore enacted . . . that it shall and may be lawful, upon complaint made by the churchwardens or overseers of the poor of any parish to any justice of peace within forty days after any such person or persons coming so to settle as aforesaid in any tenement under the yearly value of ten pounds, for any two justices of the peace . . . of the division where any person or persons that are likely to be chargeable to the parish shall come to inhabit, by their warrant to remove and convey such person or persons to such parish where he or they were last legally settled, either as a native, householder, sojourner, apprentice or servant, for the space of forty days at the least, unless he or they give sufficient security for the discharge of the said parish, to be allowed by the said justices. . . .

Provided also, that . . . it shall and may be lawful for any person or persons to go into any county, parish, or place to work in time of harvest, or at any time to work at any other work, so that he or they carry with him or them a certificate from the minister of the parish and one of the churchwardens and one of the overseers for the poor for the said year, that he or they have a dwelling-house or place in which he or they inhabit. . . .

And for the further redress of the mischiefs intended to be hereby

[89] Raw materials, such as hemp or flax, provided to afford work for the poor.

remedied, be it enacted . . . that from thenceforth there be and shall be one or more corporation or corporations workhouse or workhouses within the cities of London and Westminster, and within the boroughs, towns and places of the county of Middlesex and Surrey. . . .

210. Population and Wealth

Although there was no national census in England until 1801, and statistics with regard to the national wealth leave much to be desired, some impressions may be gained through Gregory King's estimate of the numbers and incomes of different social classes in 1688, which was published a few years later. King sets the English population at 5,500,520 and the annual national income at £43,505,800.

Number of families	Rank, degrees, titles and qualifications	Heads per family	Number of persons	Yearly income per family £
160	Temporal lords	40	6,400	2,800
26	Spiritual lords	20	520	1,300
800	Baronets	16	12,800	880
600	Knights	13	7,800	650
3,000	Esquires	10	30,000	450
12,000	Gentlemen	8	96,000	280
5,000	Persons in greater offices	8	40,000	240
5,000	Persons in lesser offices	6	30,000	120
2,000	Merchants and traders by sea	8	16,000	400
8,000	Merchants and traders by land	6	48,000	200
10,000	Persons in the law	7	70,000	140
2,000	Eminent clergymen	6	12,000	60
8,000	Lesser clergymen	5	40,000	45
40,000	Freeholders of the better sort	7	280,000	84
140,000	Freeholders of the lesser sort	5	700,000	50
150,000	Farmers	5	750,000	44
16,000	Persons in sciences and liberal arts	5	80,000	60
40,000	Shopkeepers and tradesmen	4½	180,000	45
60,000	Artisans and handicrafts	4	240,000	40
5,000	Naval officers	4	20,000	80
4,000	Military officers	4	16,000	60
50,000	Common seamen	3	150,000	20
364,000	Laboring people and out-servants	3½	1,275,000	15
400,000	Cottagers and paupers	3¼	1,300,000	6½
35,000	Common soldiers	2	70,000	14
—	Vagrants	–	30,000	—

SOURCE: Gregory King, "Natural and Political Observations and Conclusions upon the State and Condition of England, 1696," George Chalmers, *An Estimate of the Comparative Strength of Great Britain* (London: J. Stockdale, 1802), 424–25.

SECTION P

The Threshold of the Age of Reason

211. Francis Bacon on the Inductive Method

The seventeeth century witnessed an increasing interest in scientific experimentation. Sir Francis Bacon sought to create a new philosophical system, based on the inductive method. His *Novum Organum* (1620), from which these extracts are taken, sets forth the means by which "the intellect may be raised and exalted, and made capable of overcoming the difficulties and obscurities of nature."

The art which I introduce with this view (which I call "interpretation of nature") is a kind of logic, though the difference between it and the ordinary logic is great; indeed immense. For the ordinary logic professes to contrive and prepare helps and guards for the understanding as mine does, and in this one point they agree. But mine differs from it in three points especially; viz., in the end aimed at; in the order of demonstration; and in the starting point of the inquiry.

For the end which this science of mine proposes is the invention not of arguments but of arts, not of things in accordance with principles but of principles themselves, not of probable reasons but of designations and directions for works. And as the intention is different, so accordingly is the effect, the effect of the one being to overcome an opponent in argument, of the other to command nature in action.

In accordance with this end is also the nature and order of the demonstrations. For in the ordinary logic almost all the work is spent about the syllogism. Of induction the logicians seem hardly to have taken any serious thought, but they pass it by with a slight notice, and hasten on to the formulae of disputation. I on the contrary reject demonstration by syllogism, as acting too confusedly, and letting nature slip out of its hands. . . . Although . . . I leave to the syllogism and these famous and boasted modes of demonstration their jurisdiction over popular arts and such as are matter of opinion (in which department I leave all as it is), yet in dealing with the nature of things I use induction throughout, and that in the minor propositions as well as the major. For I consider induction to be that form of demonstration which upholds the sense, and closes

---

SOURCE: Richard F. Jones (ed.), *Francis Bacon: Essays, Advancement of Learning, New Atlantis and Other Pieces* (New York: Odyssey Press, 1937), 254–56.

with nature, and comes to the very brink of operation, if it does not actually deal with it.

Hence it follows that the order of demonstration is likewise inverted. For hitherto the proceeding has been to fly at once from the sense and particulars up to the most general propositions, as certain fixed poles for the argument to turn upon, and from these to derive the rest by middle terms, a short way, no doubt, but precipitate, and one which will never lead to nature, though it offers an easy and ready way to disputation. Now, my plan is to proceed regularly and gradually from one axiom to another, so that the most general are not reached till the last, but then when you do come to them, you find them to be not empty notions but well defined, and such as nature would really recognize as her first principles, and such as lie at the heart and marrow of things.

But the greatest change I introduce is in the form itself of induction and the judgment made thereby. For the induction of which the logicians speak, which proceeds by simple enumeration, is a puerile thing; concludes at hazard; is always liable to be upset by a contradictory instance; takes into account only what is known and ordinary; and leads to no result.

Now what the sciences stand in need of is a form of induction which shall analyze experience and take it to pieces, and by a due process of exclusion and rejection lead to an inevitable conclusion. And if that ordinary mode of judgment practised by the logicians was so laborious, and found exercise for such great wits, how much more labor must we be prepared to bestow upon this other, which is extracted not merely out of the depths of the mind but out of the very bowels of nature. . . .

212. The Royal Society

Notable in the development of scientific interest was the formal establishment, in 1662, of the Royal Society. It consisted of men, for the most part amateurs, engaged in scientific experimentation and the accumulation of scientific knowledge. Their objectives were summarized in 1667 as follows, by Thomas Sprat, one of the first Fellows of the Society.

I will here, in the first place, contract into few words the whole sum of their resolutions . . . Their purpose is, in short, to make faithful records of all the works of nature or art which can come within their reach, that so the present age, and posterity, may be able to put a mark on the errors

---

SOURCE: Thomas Sprat, *History of the Royal Society* (London: T. R. for J. Martyn and J. Allestry, 1667), sec. v.

which have been strengthened by long prescription; to restore the truths that have lain neglected; to push on those which are already known to more various uses; and to make the way more passable to what remains unrevealed. This is the compass of their design. And to accomplish this, they have endeavored to separate the knowledge of nature from the colors of rhetoric, the devices of fancy or the delightful deceit of fables. They have labored to enlarge it, from being confined to the custody of a few, or from servitude to private interests. They have striven to preserve it from being over-pressed by a confused heap of vain and useless particulars, or from being straitened and bounded too much up by general doctrines. They have tried to put it into a condition of perpetual increasing, by settling an inviolable correspondence between the hand and the brain. They have studied to make it not only an enterprise of one season, or of some lucky opportunity, but a business of time: a steady, a lasting, a popular, an uninterrupted work. They have attempted to free it from the artifice and humors and passions of sects, to render it an instrument whereby mankind may obtain a dominion over things, and not only over one another's judgments. And lastly, they have begun to establish these reformations in philosophy, not so much by any solemnity of laws or ostentation of ceremonies as by solid practice and examples; not by a glorious pomp of words, but by the silent, effectual and unanswerable arguments of real productions. . . .

213. Reason and Religion

The growing interest in scientific experimentation went hand in hand with a growing confidence in human reason as a means to unravel the mysteries of the universe. Religious thought was considerably affected. In "The Work of Reason," Benjamin Whichcote (d. 1683), one of a group of clergymen known as the Cambridge Platonists, emphasized the importance of reason in theology.

That which is the height and excellency of humane nature, viz., our reason, is not laid aside nor discharged, much less is it confounded, by any of the materials of religion, but awakened, excited, employed, directed and improved by it. For the mind and understanding of man is that faculty whereby man is made capable of God, and apprehensive of him, receptive from him, and able to make returns upon him and acknowledgments to him. Bring that with you, or else you are not capable receivers.

SOURCE: Benjamin Whichcote, "The Work of Reason," E. T. Campagnac, *The Cambridge Platonists* (Oxford: The Clarendon Press, 1901), 51, 53–55.

Unless you drink in these moral principles, unless you do receive them by reason, the reason of things by the reason of your mind, your religion is but shallow and superficial. . . .

I say, if . . . a man doth not admit what he receives with satisfaction to the reason of his mind, he doth not receive it as an intelligent agent, but he receives it as a vessel receives water; he is *continens* rather than *recipiens*.[90] . . .

Of all impotencies in the world, credulity in religion is the greatest. This Solomon hath observed, that simple, weak, shallow heads are foolish, and believe that which anyone saith, sail with every wind that blows. . . . When a man hath made a deliberate act of judgment in a case, upon consideration of reason, grounds and principles, he hath always ever after within him whereby to encourage him to go on and answer all objections as they shall arise. . . .

Man is not at all settled or confirmed in his religion until his religion is the self-same with the reason of his mind; that when he thinks he speaks reason, he speaks religion, or when he speaks religiously, he speaks reasonably; and his religion and reason is mingled together; they pass into one principle; they are no more two, but one: just as the light in the air makes one illuminated sphere, so reason and religion in the subject are one principle. . . .

[90] Containing rather than receiving.

G H I J K L M N O P 9 8 7 6 5 4 3